Quality of Life in Jewish Bioethics

Quality of Life in Jewish Bioethics

Edited by
Noam J. Zohar

LEXINGTON BOOKS

A division of
ROWMAN & LITTLEFIELD PUBLISHERS, INC.
Lanham • Boulder • New York • Toronto • Oxford

LEXINGTON BOOKS

A division of Rowman & Littlefield Publishers, Inc.
A wholly owned subsidiary of The Rowman & Littlefield Publishing Group, Inc.
4501 Forbes Boulevard, Suite 200
Lanham, MD 20706

PO Box 317
Oxford
OX2 9RU, UK

British Library Cataloguing in Publication Information Available

Library of Congress Cataloging-in-Publication Data

Quality of life in Jewish bioethics / editor, Noam J. Zohar.
p.; cm.
Includes bibliographical references and index.
ISBN 0-7391-1445-X (cloth : alk. paper) — ISBN 0-7391-1446-8 (pbk. : alk. paper)
ISBN 978-0-7391-1445-2 ISBN 978-0-7391-1446-9
1. Bioethics—Religious aspects—Judaism. 2. Quality of life. 3. Medicine—Religious aspects–Judaism. 4. Jewish ethics. I. Zohar, No'am.
R725.57.Q35 2005
174'.957—dc22 2005029443

Printed in the United States of America

♾™ The paper used in this publication meets the minimum requirements of American National Standard for Information Sciences—Permanence of Paper for Printed Library Materials, ANSI/NISO Z39.48–1992.

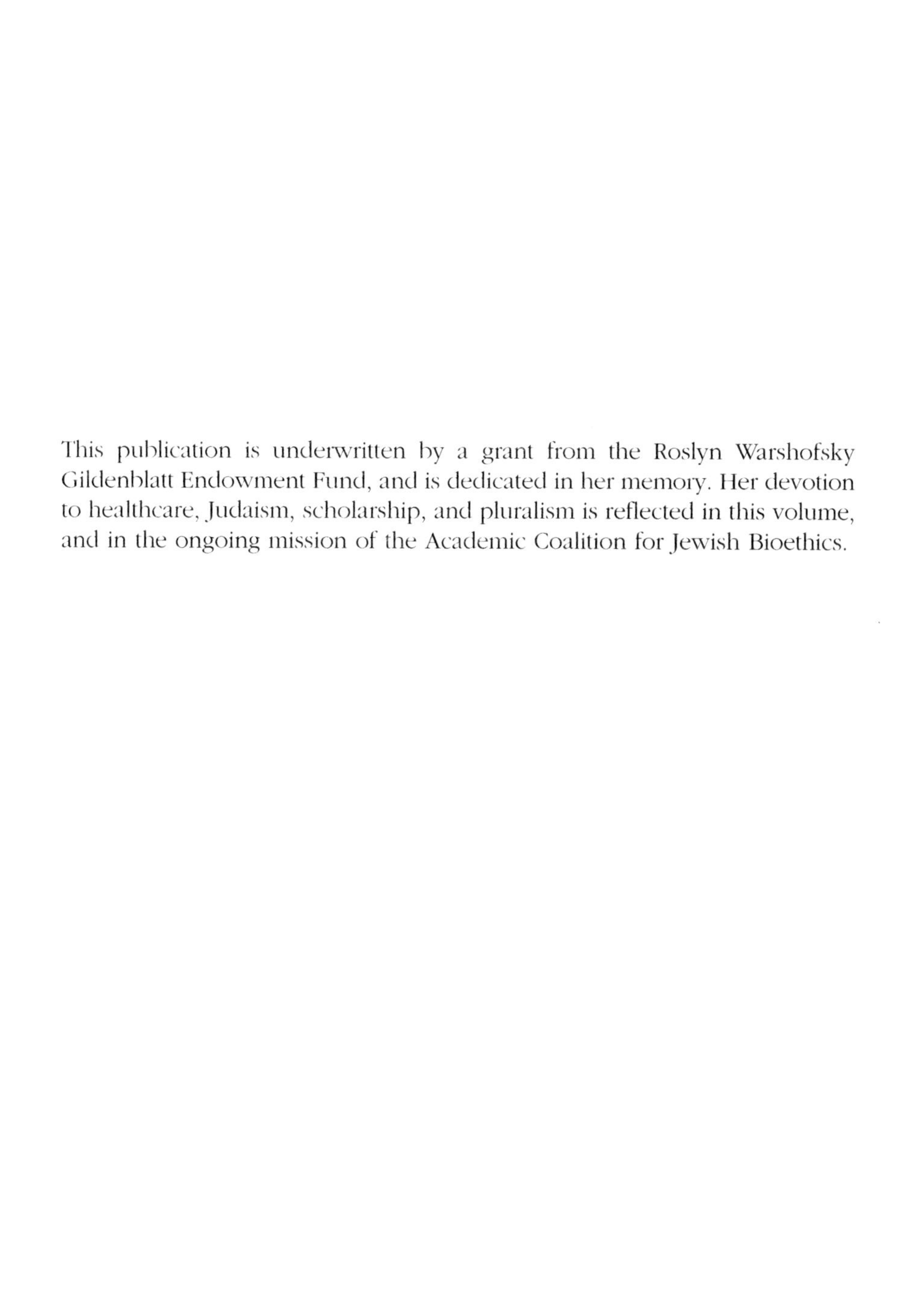

This publication is underwritten by a grant from the Roslyn Warshofsky Gildenblatt Endowment Fund, and is dedicated in her memory. Her devotion to healthcare, Judaism, scholarship, and pluralism is reflected in this volume, and in the ongoing mission of the Academic Coalition for Jewish Bioethics.

Contents

Preface

In the fall of 1999, the founding father of modern Jewish bioethics, Rabbi Lord Immanuel Jakobovits, addressed a gathering at Boston University. Many of us in the audience—Jews (and some non-Jews) of diverse persuasions and with far from uniform commitments—were deeply touched by his wisdom, but even more so by his sentiment, as he looked back on the decades that had passed since the publication of his *Jewish Medical Ethics*. We could not know that this was the last time we would hear him speak in public: Rabbi Jakobovits—may the remembrance of the righteous be a blessing—passed away a few weeks later. Even so, his words had the ring of a testament, of an elder defining his legacy and advising those who follow in his footsteps.

It was no accident that Rabbi Jakobovits was speaking to such a diverse audience. This was the second of three meetings organized by Professor Paul Wolpe, who sought to provide a framework for pluralist discourse in Jewish bioethics. Until then, much of the public voices and forums in this field had been geared rather strictly to particular "denominations"—more often than not, exclusively Orthodox. These meetings, instead, were to be guided by the pluralist spirit of classical Rabbinic teaching, *elu va-elu*: "These and those are the words of the living God." Rabbi Jakobovits welcomed the enterprise, and made it explicit that he regarded with satisfaction the diversity and wide-ranging disagreements within this field that he had effectively founded—even while disagreeing with more than one of our statements and positions.

The third and final of these meetings took place in New York in the spring of 2002. In its final session, those present voiced excitement at the unique quality of what had transpired and a shared sense of how important it was to keep this kind of discussion going. The diverse Jewish individuals and groups involved in healthcare and its ethical dimension as patients, caregivers, advisors, and teachers need a forum for sharing and learning from each other and from our rich traditions—a forum dedicated to promoting pluralistic Jewish bioethics, open to many sensibilities and understandings. This is the mission undertaken by the Academic Coalition for Jewish Bioethics (ACJB)—a coalition of academic institutions committed to promoting Jewish bioethics, each out of its own faith commitments, and jointly out of a commitment to pluralistic discourse and mutual respect. This volume is the first published fruit of that coalition.

The first ACJB conference took place in Philadelphia in the spring of 2004. The organizers believed that the best way to inaugurate the hoped-for vigorous discussion would be to take on an issue that had heretofore been almost taboo in Jewish bioethics—not because it lacked importance, but because it had been perceived as somehow incompatible with fundamental Jewish values: the issue of "Quality of Life." If there is one value-emphasis that seems undisputable and sacrosanct in Jewish bioethics, it is the value of human life itself. "Life saving takes precedence over everything"—has been a familiar Judaic motto; and in the context of medical ethics, this has commonly been taken to preclude any consideration of "quality of life." Yet such considerations cannot, in the end, be ignored. Thus the theme of ACJB's first conference was announced: "Addressing Quality of Life: A Challenge for Jewish Bioethics." The presentations made at that inaugural meeting formed the basis for this book. We hope that this will be the first of a series of publications promoting an inclusive and committed discussion and understanding of Jewish bioethics.

My work in editing this volume was warmly supported by the leadership of the Shalom Hartman Institute in Jerusalem, and for this I wish to register my deep appreciation to David and Donniel Hartman and to Yehudit Schweig. My work on this project began during my year of membership in the "bioethics group" at the School of Social Science in the Institute for Advanced Study, Princeton. The project would never have succeeded without the continual help and support of Donna Kirshbaum, the first ACJB coordinator, and of ACJB founding president David Teutsch.

NJZ

Abbreviations and Mode of Citation

Most rabbinic texts have been reproduced in many forms, printed and digital. Therefore, references to these texts are not to any particular edition; instead (following common practice in Judaic scholarship), citations point to the standard divisions and subdivisions. These may be chapter and verse (as in the Bible), folio and side (as in the Talmud), or section/subsection (as in various rabbinic works). We have used the following abbreviations:

BT Babylonian Talmud
JT Jerusalem Talmud (also known as the "Palestinian Talmud")
MT Mishneh Torah (known in its English translation as The Code of Maimonides)
SA Shulhan Arukh (the sixteenth-century code by Rabbi Joseph Caro)

And its sections:
EH Even ha-Ezer
HM Hoshen Mishpat
OH Orah Hayyim
YD Yoreh De'ah

Note: Definitions for some terms may be found in the notes; these can be located by means of the index.

THE BOARD OF THE ACADEMIC COALITION FOR JEWISH BIOETHICS

THE WORK OF THE ACJB HAS BEEN MADE POSSIBLE BY THE GENEROUS SUPPORT OF:

Gimprich Foundation
Kalsman Institute on Judaism and Health
Daniel Levin
Joseph Lieb Memorial Fund
Levin-Lieber Program in Jewish Ethics
Maurice Amado Foundation
Roslyn W. Gildenblatt Memorial Endowment Fund
The Solomon and Sylvia Bronstein Foundation
Whizin Support Foundation

Introduction: The Expanding Goals of Medicine and the Quest for Refocusing Jewish Bioethics

Noam J. Zohar

Popular television shows about hospitals tend to focus on emergencies and (often averted) catastrophes, thus helping to maintain, in our minds and popular culture, the association of medicine with the high drama of life-and-death. Even as the goals of medicine in the developed world continue to expand, with growing attention given to promoting quality of life rather than just saving life as such, a powerful image persists of the health-care team in green and white coats rushing to stave off death.

Yet thinking about quality of life is inseparable from the project of life-saving. Facing hard choices, patients sometimes contemplate whether to opt for a desperate procedure, risking an outcome that they might regard as worse than death.[1] No less poignantly, providers must constantly make decisions—on both the micro and macro levels—about priorities in the allocation of efforts and resources, and are naturally led to assess the relative benefits to their actual or prospective patients. In addressing these quandaries, much effort has been directed toward defining *measurements* for quality of life; these efforts are described and explained (and, in part, criticized) by Paul Wolpe in our opening chapter. In this context it is worth mentioning a key term that has not won a place in the discourses of Jewish bioethics (and accordingly makes no appearance in this book's chapters), namely Quality Adjusted Life-Years, or QALYs.[2]

Moreover, "quality of life" is by no means only a vital dimension in thinking about when and how to preserve patients' lives. It also signifies an independent goal—or rather, a broad and expanding range of goals—of medical treatment and interventions. Yet within the ambit of Jewish bioethics, there has been a noticeable reluctance, or even opposition, to talking about quality of life. The generally prevalent image of medicine as primarily life-saving is combined, in the Jewish context, with an extraordinarily strong traditional emphasis on life-saving as a supreme imperative and on preserving human life as an almost absolute value. The classical talmudic discussion (in the last chapter of tractate *Yoma*) starts from the permission—nay, requirement—to break one's fast on the Day of Atonement if there is even a slight danger of death. From that, it leads to an elaborate set of justifications for desecrating the *shabbat* for the sake of life-saving; not surprisingly, this section of the Talmud is cited in more than one of the essays in this volume.

The principle, "Life-saving overrides the [sanctity of] *shabbat*"—and by obvious implication, virtually any other *mitzvah*—is a commonplace in traditional Jewish society. Hence in Jewish bioethics, even more than elsewhere, the value of medicine has been defined primarily in terms of life-saving, and in turn the preservation of human life has been seen as the central and overriding value within medical practice. Concerns about "quality of life" are often regarded with suspicion, as they seem to compromise our commitment to life-saving.

The reluctance does not stem, however, only from the perceived conflict with life-saving. The title of Laurie Zoloth's essay (chapter 3) juxtaposes "Qualities of Mercy" and "Quality of Life," suggesting that in wondering about whether it is worthwhile to save someone's life, questions are raised about the moral qualities of the hesitating agent. Echoing the evocative title that Maimonides gave to a pertinent section of his Code ("The laws of Murder and the Protection of Life"—*shemirat nefesh*), Zoloth (drawing on the work of Emmanuel Levinas and Benjamin Freedman, among others) develops a religious ethic of responsibility, of the caretaker's duty (*shomer*) to protect and safeguard.

Thus claiming a significant place for "quality of life" within Jewish bioethics, as we do in this book, constitutes an almost fundamental challenge to the prevailing paradigm in this field. This challenge—or rather, this constructive criticism—is supported by the volume as a whole, which promotes recognition of the many-sided values represented by "quality of life." The opposition deriving from the prevailing paradigm is addressed directly in my own contribution (chapter 2), where I ask whether, in traditional Jewish terms, a quest for promoting quality of life is a good thing. The answer is not uniform, since besides a tradition (arguably, the mainstream) affirming the worldly and the human body, there is also a Jewish ascetic tradition, which may well look askance at a pursuit of life's enjoyments.

Now, even if—rejecting asceticism—one looks positively upon promoting quality of life, this would seem to necessarily take a backseat as compared to life-saving. Depending on just what is meant by "quality of life" (noting its ambiguities, as analyzed by Wolpe), a Judaic case can nevertheless be made—or so I argue—for according it significant value.

Our understanding of what "quality of life" is about is enriched by William Cutter's narrative perspective. Cutter shows (in chapter 4) how misleading it can be to perceive quests for (medical) enhancement in snapshot mode, failing to contextualize them in the individual's life-story. It is the personal narrative, as well as the cultural narrative prevalent in a particular society, that illuminates the significance of particular preferences and efforts. Likewise, citing the philosophical tradition of narrative ethics, Cutter shows (drawing on the work of Robert Cover) how halakhic teachings and traditional normative judgments are embedded in literary narratives, from the aggadic stories in the Talmud to the tales of Rabbi Nahman of Bratzlav.[3]

Part one concludes with Elliot Dorff's discussion of hope and acceptance, particularly at the end of life. Presented as a review essay of a book by Jerome Groopman (2004), Dorff's analysis (chapter 5) illuminates the relationship between "quality of life" and life-prolonging interventions from the special perspective of hospice care. As Dorff reports, the late Rabbi Lord Immanuel Jakobovits expressed concern over the philosophy of the first Jewish hospice in Los Angeles, echoing the suspicions described above: Does not the hospice emphasis on "quality of life," on palliation and simple caring, involve "abandonment of all hope for recovery"? Dorff replies that Judaic teaching does not encourage unrealistic hope; here the time for heroic life-saving medicine is already long past. Yet the patient can hope and pray for painless and meaningful months or weeks—for quality of life in the time that remains.

A striking element of Dorff's exposition is his emphasis on the crucial role of personal interactions, put forward in terms of the *mitzvah* that is central in this context—visiting the sick. For all patients, but most especially for those under the shadow of death, the quality of their days and hours is greatly enhanced by interpersonal connections. Note, however, that these caring relationships are enacted within the timeframe of a visit, or a series of visits. The story becomes very different when close relationships—first and foremost, family relationships—become the nexus of long-term care. Here the tension between prolonging life and living well takes a wrenching turn, for the burden of care—grounded in love and commitment—may take an enormous toll on the caregiver. This is the subject of the two chapters that comprise part two.

In this part of our exploration, modern medicine with its great capabilities is not the means for attaining a (putative) improvement in quality of life, but rather is itself producing a serious threat to people's quality of life, through the demands it makes on caregivers within the family. Dayle Friedman reminds us that moral quandaries are not encountered in a vacuum, but rather arise in particular historical and social settings. Medical advances, together with other factors, have produced a great increase in longevity. As Friedman warns us, this "age wave" "creates a parallel *caregiving* wave. Jewish families are caring for elderly members living longer, and with more extended periods of greater dependency, than ever before." (Interestingly, this same demographic feature figures in a rather different way in chapter 8.)

Friedman lays out the basic dilemma starkly, by citing a collection of hyperbolic statements and stories about exemplary filial piety. These talmudic models appear to recognize no limit to children's duties under the fifth commandment, "honor your father and mother." Parental frailty, especially in combination with healthcare-related needs, seems to leave scarce room for considering the caregiver's quality of life. Against this, Friedman cites some clear halakhic caveats, but more importantly, she makes the crucial observation that relationships and their concomitant obligations are inherently reciprocal, and continue overall to be so. The analysis here is couched primarily in terms of values, and these include the value of the (adult) child's own life, dreams, and commitments. Friedman reminds us that at the founding moment of Judaism, Abraham left his father behind to go forth in response to God's calling.

In the next essay (chapter 7), Deena Zimmerman addresses many of the same issues, and these two chapters—which comprise part two of this volume—can be read together, as an illuminating dialogue. Both draw, for example, on the classical talmudic discussion of filial and familial duties (in tractate *Kiddushin*)—though with striking differences in emphasis and interpretation. Also, both note poignantly the gender gap: the tasks of caregiving within the family fall chiefly upon women. Zimmerman engages in some detailed halakhic analysis, which facilitates a clear distinction between monetary obligations of care—largely subsumed under *tzedakah*—and duties of personal care and involvement. It is the latter that significantly impact the quality of life of both parties in the caregiving relationship. Among those carrying greater burdens, Zimmerman points our attention to parents caring for children with special needs—and notes the crucial importance of community attitudes. These observations are instructive in showing that a concern for people's quality of life extends far beyond the restricted set of instances where it might compete with life-saving.

Indeed, within the goals of medicine itself there has been a growing inclusion of efforts and enterprises directed solely toward improving patients' quality of life. Alleviation of pain has always been proclaimed as an important part of the healer's role,[4] but more recently mainstream medicine has increasingly embraced application of medical science and technology toward a broad range of life-quality enhancements. Does broadening the scope of "medicine" imply a parallel broadening of the concepts of "illness" and "health"—or should we instead recognize that medicine need not be restricted to combating illness or to restoring health?

This question has led to a proposed distinction—one much debated—between "therapy" and "enhancement." Among the many proponents of this distinction there is a fundamental divergence regarding the nature of the conclusions they wish to draw from it. Some mean to re-affirm therapy as the sole, or chief, goal of medicine, relegating most forms of "enhancement" to the general marketplace of goods and services (and some of its forms to the realm of the forbidden). Others, on the contrary, wish to explicitly endorse enhancement, side by side with therapy, as central to the enterprise of medicine. And then there are those who altogether deny the coherence of the distinction itself.

"Enhancement" is aimed, by definition, not at life-saving but at improving the quality of life, and is thus appropriately the subject of the last part of this volume. Part three opens with an essay written by Thomas Cole, together with Robin Solomon, whose first section provides a concise survey of the literature on "Enhancement versus Therapy." This is then applied to purported promises of "anti-aging medicine." Is aging itself a disease? If not, then perhaps coping with it—in a quest for a better quality of life—does not call primarily for medical interventions, but rather for a "persuasive vision of the good life for our elder years," which (the authors maintain) is sadly lacking in contemporary Judaism.

A Judaic version of the normative debate over enhancement is represented in our two concluding chapters. It is very instructive that both cite the same classical text (each uses a different translation; and each, of course, cites other sources as well), yet reach almost opposite conclusions. Focusing on the issue of cosmetic surgery, Mordechai Halperin expresses a basically liberal halakhic stance, positing a permissive presumption: anything for which no prohibition has been established is by default permitted. Here Halperin addresses the halakhic prohibition of self-harm (a theme that is also explored, in another context, in chapter 2), which might be adduced as a basis for forbidding surgery that is not medically required. He concludes (citing Rabbi Moshe Feinstein) that if an intervention is deemed to be objectively beneficial, and is also truly desired by the individual in question, then it is not included in the ambit of the prohibition.

Halperin also locates his discussion within the larger theological debate about the permissibility of medical intervention, in light of belief in divine Creation and Providence. It is here that he cites a well-known *midrash* that compares the work of a physician to that of a farmer. Agriculture, like medicine, regards Nature—as given by God—as inherently requiring improvement. Louis Newman, in chapter 10, cites the same *midrash*—whose theological stance he understands no differently than does Halperin—yet prefers a more restrictive interpretation. Newman's discussion is aimed at the question of enhancement more broadly; in particular, he has in mind prospects for genetic enhancement. In his analysis, he contrasts two value orientations: the pro-interventionist (through the aforementioned *midrash*) versus a stance that consciously refrains from assessing some of God's creations as less valuable than others. In support of this, he cites the talmudically ordained "blessing," or utterance of praise to the Creator, to be recited upon encountering difference—a source that was adduced in chapter 7 as a hoped-for inspiration for communal acceptance of children with special needs. This connection brings to mind a recurring theme in debates about enhancement: Our quest for improvement should not lead us to devalue all those—actually, each and every one of us—who are less than perfect.

The essays in these three parts do not, of course, exhaust the topic of "quality of life" in bioethics, or even specifically in Jewish bioethics. But it is my hope that they offer a rich and robust sense of what might be included under this theme, and thereby establish for it a serious place in the discourse of Jewish bioethics, alongside that rightly accorded to the questions pertaining to life-saving.

These ten essays vary greatly in their methodologies, probably far more than is common in other books of this genre. This is of course no accident—it derives from, and gives expression to, the pluralist philosophy that guided the coming together of these scholars in the first place. It is thus perhaps worthwhile to point out here some aspects of this methodological variety. This is no substitute, of course, for a systematic exploration of method in Jewish bioethics—a task that has been admirably begun by others and also (to a small extent) by myself in earlier work.[5] I shall not even attempt a thorough second-order analysis of the arguments presented in this volume, as that would far exceed the proper role and scope of an introduction. Instead, I shall merely indicate a few points of interest.

The Jews have been called "People of the Book"; and although this originally signified the Torah in its narrow sense—the Scroll including the five books of Moses—it is also true regarding the Torah in its broader sense, which includes all valid Judaic teachings from Sinai down to the present. Consequently, Judaic method is text-oriented. Most of our authors cite texts extensively, and all do so to some degree. Interestingly, in several instances the same text—or texts excerpted from the same pages of the Talmud—are cited in more than one chapter. Sometimes this is done in surprisingly diverse contexts, but sometimes the reader is treated to alternative analyses or applications of the same text to a contested issue (reminiscent of the sages of old). I noted above two striking examples of this—the Friedman-Zimmerman discussion in chapters 6 and 7, and the Halperin-Newman debate in chapters 9 and 10; and there are several others. Recourse to a common textual tradition obviously does not ensure agreement; this is particularly true in this enterprise, since our authors represent a broad spectrum of attitudes toward traditional texts and their authority.

It is worth noting that a stronger commitment to *halakhah* does not necessarily translate into endorsement of more restrictive positions. Nor does it exclude recourse to non-halakhic sources or to ethical considerations in their own right, as can readily be seen by examining the essays in this volume composed in a primarily halakhic mode of discourse (those by Halperin, Zimmerman, and—in part—Dorff and Zohar). Nor, conversely, do other modes of ethical discourse preclude an engagement with halakhic sources. This is evident in the essays by Friedman and Newman, whose explicit framework is that of value analysis; and equally in that by Zoloth, which might be characterized as talmudic-theological.

For several of our authors, consciousness of historical change and setting is a vital component of bioethics. This is most prominent in Cutter's "narrative ethics" discourse, which significantly includes a history (or at least, a story) of the evolution of narrative ethics. This approach instructs us to find the meaning of ethical questions in their social and historical context—and to turn to the tradition as a source of illuminating stories, which crucially give meaning even to norms. But contemporary Jewish history figures also in the presentations by Cole and Solomon and by Friedman, which both stress the significance of the "age wave." It is also a source to which Zimmerman turns where she finds little else to go by—inferring Jewish values (regarding communal acceptance of people with disabilities) from the historical record—gleaned, in this case, from halakhic sources.

Historical exposition also plays a key role in Wolpe's opening chapter, where our theme concept—"quality of life"—is introduced through a critical historical account. Without this kind of account, our Judaic explorations could not have proceeded (a similar observation applies to the exposition by Cole and Solomon at the beginning of part three). This is true, however, not only with respect to history, but also with respect to philosophical bioethics generally. For all of our authors—albeit to differing degrees—the enterprise of Jewish bioethics is vitally informed by the concepts and arguments of the several disciplines that come together in the field of bioethics.

I believe it is equally true that the broader discourse of bioethics has much to gain from listening to the voices and insights of particular traditions. This is especially so for any theme that stands in tension with some prominent value of a given tradition. Precisely because Judaism has always emphasized life-saving above (almost) all, its grappling with the concept of "quality of life" is uniquely illuminating.

REFERENCES

Dworkin, Ronald (1993), *Life's Dominion: An Argument about Abortion and Eutanasia*, New York: HarperCollins.

Groopman, Jerome (2004), *The Anatomy of Hope: How People Prevail in the Face of Illness*, New York: Random House.

Mackler, Aaron (1995), "Cases and Principles in Jewish Bioethics: Toward a Holistic Model," in *Contemporary Jewish Ethics and Morality*, ed. Elliot N. Dorff and Louis E. Newman, New York: Oxford University Press, 177–93.

Nord, Erik (1999), *Cost-Value Analysis in Health Care: Making Sense out of QALYS*, New York: Cambridge University Press.

Randall, Fiona, and Downie, R. S. (1996), *Palliative Care Ethics*, New York: Oxford University Press.

Zohar, Noam J. (1997), *Alternatives in Jewish Bioethics*, Albany: SUNY Press.

NOTES

1. For an account of how such a preference can make sense, even if the outcome at stake is an existence without consciousness (wherein the individual supposedly no longer has any interests or concerns), see Ronald Dworkin's discussion of "critical interests" (Dworkin 1993, 199–217).

2. For a lucid exposition and constructive critique, see Nord (1999).

3. *Halakhah* is the Jewish tradition of normative discourse, sometimes called more narrowly "Jewish Law." *Aggadah* denotes the vast non-*halakhic* parts of the rabbinic tradition, including homilies, non-legal biblical interpretations, parables and tales from the lives of the sages, and more.

4. Proclaimed—but not necessarily always followed in practice; see Randall and Downie, 1996.

5. See especially the broad-ranging analysis by Aaron Mackler (1995), which includes numerous references to efforts by others; and see my introduction in Zohar (1997), 1–16.

I

QUALITY OF LIFE VS. SANCTITY OF LIFE IN CONTEMPORARY AND JUDAIC DISCOURSE

1

Understanding "Quality of Life": Does It Have a Role in Medical Decision-Making?

Paul Root Wolpe

INTRODUCTION

Quality of life is an elusive concept; we think we know what it means until we actually try to define it. Yet not only is the idea used in almost all areas of medicine; we make some of our most important life and death decisions based on a patient's presumed quality of life. Over the last few decades we have abandoned the idea that medicine is obligated to maintain life at all cost, and have recognized that, in certain circumstances, other values may be more compelling than life for life's sake alone. We have also begun to demand easy-to-quantify measures to determine what to pay for. We have therefore begun to ask ourselves: What are the factors that we should consider in end-of-life decision making? Is quality of life one of those factors? And if so, what exactly is it, and does it mean different things to different people or in different circumstances? What are the components of quality of life that insurance companies should pay to help us regain after illness or injury? These questions, and others, demand a robust and clear definition of quality of life, and a reliable way to measure it. Yet the concept remains a difficult one to pin down.

It may come as a surprise to many readers that "quality of life" is a relatively new concept. The idea emerged in the professional literature only about twenty-five years ago, and a few years before that quality of life as a concept is almost absent. When a concept arises so quickly and immediately to assume an important role in medical decision-making, it is instructive to explore why the term emerged in that particular historical and cultural moment.

How was medicine changing when quality-of-life measures were first proposed? One of the most dramatic changes was our ability to maintain people for long periods on life support, especially at the end of life. Questions began to arise about what kinds of treatments were appropriate for dying people or people destined to live out the remainder of their lives in a persistent vegetative state. So even though the idea of "quality of life" has now transcended its original meaning and is used in a variety of other medical (and even non-medical) contexts, it is worth noting that "quality of life" started out as an attempt to try to solve some clear and practical problems of end-of-life care. Today we have literally thousands of articles that come out every year on various medical, economic, social, ethical, religious, and other aspects of quality of life. In fact, there is nothing intrinsically medical about the idea itself, and the term could just as easily refer to the innumerable non-medical aspects of our enjoyment of life—our wealth, environment, whether we have access to cultural institutions, our job stress, and so on. Many things contribute to our quality of life. My discussion here will, however, be restricted to quality of life as it relates to medical issues and medical decision-making, and not include those people who want to improve their quality of life by moving to Bucks County and starting a llama farm.

Why do we need to discuss quality of life at all? I think there are two main reasons in the medical context. First, quality of life is used to assess medical interventions. Such assessments may be undertaken by individual or by institutional agents: a physician may consider quality of life in deciding what kinds of medical interventions are appropriate for a particular patient, insurance companies use them to decide which therapies offer the most benefit for their dollar, medical centers must factor them in when they consider into which approaches they want to channel their research efforts and resources. Each of these requires a means to judge which of two or three or four possible interventions leads to a better outcome, which one brings the greatest benefit to the patient. So quality of life is often used to decide between medical treatments, given the outcome goals of the patient, the physician, or the institution.

The second reason we talk so much about quality of life these days is because most of us, sooner or later, face the prospect of having to make difficult medical decisions about loved ones. The lived experience of making decisions about such things as care at the end of the life cycle or what to do with a severely impaired newborn makes real the moral conundrum of deciding when to intervene and when not to intervene. Knowing our loved ones as we do, understanding their fears, desires, hopes, values, beliefs, religious obligations, and so on, we often find that quality-of-life considerations become a crucial part of medical decision-making.

Below, we will look at the ways physicians and medical institutions think about quality of life, and why developing reliable measures is so difficult and so important. In doing so, perhaps surprisingly, we will also shed light on the ways we as medical decision-makers might more sensitively consider quality of life issues when making medical decisions about ourselves or those we love.

MEASURING QUALITY OF LIFE

Thousands of articles come out every year about quality of life, but most of them are dedicated to arguing with previous articles about what, in fact, quality of life actually is. Do we really know what we mean by this concept? In some sense we do; people often talk about quality of life and have a fruitful conversation without ever defining it, just assuming we are all generally discussing the same thing. All might agree that quality of life involves elements such as functioning well, enjoying life, not being in pain, maintaining our mental faculties, and living according to some set of values. Such generalities work in casual conversation. But trying to actually pin down the concept in a way that would be useful to researchers or institutional decision-makers is far more difficult.

In the literature, researchers often get around the problem of definition by uncritically accepting definitions and measures that have been used before. When they decide to perform a study that requires a quality-of-life measure, they reach for the nearest quality-of-life index or scale and use it, without actually asking the question, "What is it that I *really* want to try to figure out?" Most of the measures in the literature are known as "health related quality-of-life scales," and look primarily to selected aspects of a person's health status. Therefore, even when researchers might prefer a more general scale of a person's quality of life, they often end up with a measure that is intended to focus on health issues. The assumption behind health related quality-of-life scales is that one can separate out from the holistic nature of a person's existence certain aspects of life that are "health related," and assess the person's functioning in just those areas. The problem is that health related quality-of-life scales create an artificial distinction; as we shall see, people do not generally separate out or isolate health from their overall considerations of their quality of life. Important factors are missed when we think of health too narrowly in quality of life assessments.

Researchers have tried a number of different measures for quality of life (see, for instance, two special journal issues on the question: Ventegodt, Merrick, and

Andersen, 2003; Wade, 2003). There are scales trying to assess traits like *perceived distress* by measuring subjective diminishment in quality of life; the instrument might question a person on how upset she is about her medical condition, for example. Some scales try to measure *impact of illness*, that is, to compare a person's condition now versus her condition before she got the illness. An impact of illness approach seems to have the advantage of objectivity, which may be missing in perceived distress measures; yet impact measures have difficulties in trying to quantify pre- and post-illness quality of life. It is impossible, for example, to use impact of illness measures for someone who was born with an illness, because there is no pre-illness baseline. The *physical functioning* approach measures how well a person can execute the necessities of daily life, from simple daily life functioning (eating, dressing) to advanced functioning such as work. Other quality-of-life measures include *life-satisfaction scales*, looking at how a person has adjusted to the way things are, and *health status scales,* trying to directly assess various degrees of health and illness.

The measures above tend to have two basic features. One is that they try to assess functioning, or a person's attitude toward functioning, in relation to how a person's body used to function before illness or injury, or in relation to some presumed standard of "normal" human functioning. To do so, however, one must conceive of a person's overall sense of quality of life as made up of smaller units of quality of life (how well I used to walk, how often I used to have to take pills, what activities I used to do that I cannot do now, etc.) and then assess each and "add them up" to an overall quality-of-life measure. But are human lives made up of discrete, identifiable components of quality of life? Can quality-of-life measures really separate out these different components and measure and assess them?

Another difficulty with quality-of-life scales is the tendency to separate other aspects of the human experience from the scales on the grounds that they are not directly about "health," even though they are health related. As a result of this separation, the scales often fail to recognize the connection of health to issues such as income, work-status, professional and career disruption, interpersonal relationships, coping strategies, responsibilities, self-image, routine daily disruptions, and spiritual crises. Studies that look at these aspects of illness show a profound connection between such psychosocial processes and health status. It is very difficult to separate out "health related quality-of-life" issues and not delve deeply into reverberations of illness throughout one's web of relationships and web of psychological structures.

LIFE SATISFACTION AND QUALITY OF LIFE

In conceptualizing quality of life, researchers tend to split the world into mental and physical domains. The result is a set of life-satisfaction scales on one hand, and scales of physical functioning on the other. The operative assumption of the latter is that if you can return someone to something like normal functioning, then you can improve their quality of life. Certainly, if a person's ability to negotiate their world is impaired, they generally appreciate being returned to normal functioning. But people often report high levels of life satisfaction even in the absence of full functioning. A number of studies show high life satisfaction, strong desire to continue living, and an overall sense of joy in people who are in situations that most of us would consider to be extremely trying. A 1991 study of people with Duchenne muscular dystrophy (DMD), all of whom were on long-term mechanical ventilation, showed surprisingly low levels of dissatisfaction with life. The authors conclude "that the vast majority of severely disabled chronic ventilator-assisted individuals with DMD have a positive affect and are satisfied with life despite the physical dependence which precludes many of the activities most commonly associated with perceived quality of life for physically intact individuals" (Bach et al., 1991).

The tendency of the able-bodied to underestimate the quality of life of the sick and disabled was dramatically illustrated during the assisted suicide debate in the United States. One of the groups that was most vocal against assisted suicide was called "Not Dead Yet," a group of people with severe muscular, skeletal, and other disorders that forced most of them to live in mechanical wheelchairs. All their lives they've been told: "How could you possibly want to live that way? If I were you, I would want to be dead!" But, like any other group of people, they treasure their lives; these are the only lives they know. The members of Not Dead Yet were scared that if physician-assisted suicide became law, they would be the first people expected to use it. Yet so-called able-bodied people often seemed to assume that those who suffered from such severe conditions would be in favor of physician-assisted suicide, would be glad to have the option to end their difficult lives.

The point is not limited to people with chronic disabilities. In Oregon, where physician-assisted suicide is legal (though under strict conditions), it turns out that when physicians prescribe pharmaceuticals to dying people so they can end their life, a large percentage, around 40–50 percent, never use the drugs (Wineberg and Werth, 2003). They put the drugs in the drawer, and knowing that they're there is enough for them. It is not dying that scares them, it is the fear of loss of control in their lives. They know that if the moment comes when the pain is unbearable, they have a way out, and that is enough to allow them to die naturally.

A person may be infirm, may be old, may even be bed-ridden and in pain, and yet have a peaceful *neshamah* (soul), have some sense of joy in seeing their loved ones. They may have a quality of life that is perfectly acceptable to them at the time, and they may even report greater life satisfaction than that experienced by able-bodied people who have not yet found a sense of peace in their lives.

The same is true more generally: people whose health has deteriorated will often report the same level of quality of life that they had before their health deteriorated. We expect infirmity to diminish quality of life; some even construct quality of life scales to measure infirmity, assuming that as one's health deteriorates so must one's quality of life. Yet, people who get sick, even become bed-ridden, will often report a similar quality of life as before their illness even though their functioning has dramatically decreased.

Clearly quality of life means more than physical functioning. It includes those things that give our lives meaning and inspire in us the desire to continue living. Surprisingly, however, many quality-of-life scales, especially scales of functioning, leave out assessment of things like levels of religious and spiritual satisfaction. I consulted on an Alzheimer study one summer, and the investigators came in to show me their questionnaire. They were trying to test ways that people with Alzheimer's could set up their environments at home to maximize their ability to function in that environment, for example, by putting things they use every day always in the same place. Yet, there was nothing in their questionnaire about the subjects' religious life. What if part of their daily cadence was prayer, or other actions in their religious life? It never occurred to the team to even ask that question. What about other patterns of religious observance, what about weekly religious actions? What if you are a religious Jew and the lighting of *shabbat* candles and the celebration of *shabbat* was part of the greater sense of rhythm and structure to your life?

More recently, a number of researchers have put together scales and studies looking at spiritual and religious issues and their impact on quality of life (see, for instance, Baker, 2003; Bremer et al., 2004; Daaleman and Frey, 2004; Gall, 2004; Laubmeier, Zakowski and Bair, 2004). Such scales represent a step forward in integrating spiritual and religious needs into the overall measurement of what quality of life really means to individuals in lived relationships to their families and communities.

CONCLUSION: CONSTRUCTING A MEANINGFUL QUALITY-OF-LIFE SCALE

At the base of all these discussions of quality of life is the fundamental question, is there really an optimal level of human functioning? Does life have a single goal, or even a set of goals, against which we can assess its quality? Quality-of-life scales often assume some sort of average level of functioning that everyone should have. Yet even among able-bodied people, levels of functioning are extraordinarily varied. Many of us function with different capabilities in different contexts. If I walk onto the basketball court, my functioning is virtually at the level of disability compared to Allen Iverson. Is it not the meaning we give to life that determines our satisfaction with it, and if so, do we not need to supplement functioning with an individual sense of our life's purpose?

Instead of taking measures of people's actual functioning, suppose we took measures of how they felt about their functioning. After all, who decides what is a lower quality of life? The Talmud tells us that only a person's heart knows its own bitterness (BT *Yoma* 83a, citing Proverbs 14:10). So only each individual really knows what his or her quality of life is. We might begin by asking people simply, "How do you feel about yourself and your life?" Obviously, there are design, measurement, and other technical problems with that approach. For example, it would be hard to compare between people, because perceived wellbeing is very different between different human beings. People have different baseline levels of perceived happiness and different baseline levels of perceived satisfaction with their lives. Certainly such a question is not sufficient, but it might allow a broader set of variables around what makes life worth living in individual cases.

Perhaps we are simply trying to pack too much into the term "quality of life." It is clearly important in various contexts to know something about people's physical functioning and whether medical intervention increases that functioning; but is that really quality of life? It is simply a measure of increased ability to translate physical action into desired ends. When we ask a family to consider what a loved one's quality of life would be if we keep them on a respirator, or how good a young child's quality of life would be after an arduous liver transplant, we are asking the concept to bear a heavy moral burden. Is it up to that task? There are some children who will fight tooth and nail to stay alive; walk along the pediatric oncology ward and you will see warriors there, individuals who want to stay alive and are fighting the good fight at the age of five or six. But there are others for whom it is not worth it, for whom one more procedure or the idea of another transplant to replace a rejected organ seems insurmountably burdensome. Thus applying quality-of-life scales as a uniform measure of patient outcomes may be to miss the dynamic and important sense that each person has an inner compass about his or her own life. We need to know a person's boundaries and their sense of self before we can situate our measures of functioning and life satisfaction.

We all live with an inner biographical narrative, the story we tell ourselves about ourselves—how we got to where we are, who we are now, and who we will be in the future. We project a trajectory of self, an explicit or assumed idea of what our lives will be like ten or fifteen years from now. Rarely do relatively healthy people write into that narrative a sudden illness or a downward spiral. When illness comes and shatters that biographical narrative we are forced to rewrite it, to change the story to reflect a new status that has been thrust upon us. Perhaps, to some degree, quality-of-life measures tell us something about where we are in that process, how far down the path we have come to acceptance and incorporation of our new selves into our own inner biography. It is from within the process of rewriting our life story that a quality-of-life scale may have its most profound impact.

For those who have been ill since birth, quality-of-life measures may reflect the resilience of their spirit. For those ill after a healthy life, quality-of-life measures may reflect our ability to adapt, to accept, and then transcend the difficulties life thrusts upon us. In order to truly develop a measure of quality of life that is useful for medical decision-making, it is not enough to know how medicine and therapy have transformed the body; we must also know how healing has transformed the soul.

REFERENCES

Bach, J. R., Campagnolo, D. I., and Hoeman, S. (1991). "Life Satisfaction of Individuals with Duchenne Muscular Dystrophy Using Long-Term Mechanical Ventilatory Support." *American Journal of Physical Medicine & Rehabilitation*. 70(3): 129–35.

Baker, D. C. (2003). "Studies of the Inner Life: The Impact of Spirituality on Quality of Life." *Quality of Life Research*. 12 Suppl 1: 51–7.

Bremer, B. A., Simone, A. L., Walsh, S., Simmons, Z., and Felgoise, S. H. (2004). "Factors Supporting Quality of Life Over Time for Individuals With Amyotrophic Lateral Sclerosis: the Role of Positive Self-Perception and Religiosity." *Annals of Behavioral Medicine*. 28(2): 119–25.

Daaleman, T. P., and Frey, B. B. (2004). "The Spirituality Index of Well-Being: a New Instrument for Health-Related Quality-of-Life Research." *Annals of Family Medicine*. 2(5): 499–503.

Gall, T. L. (2004). "Relationship with God and the Quality of Life of Prostate Cancer Survivors." *Quality of Life Research*. 13(8): 1357–68.

Laubmeier, K. K., Zakowski, S. G., and Bair, J. P. (2004). "The Role of Spirituality in the Psychological Adjustment to Cancer: A Test of the Transactional Model of Stress and Coping." *International Journal of Behavioral Medicine*. 11(1): 48–55.

Ventegodt, S., Merrick, J., and Andersen, N. J. (2003). "Editorial—a new method for generic measuring of the global quality of life." *The Scientific World Journal*. 3: 946–9.

Wade, D. T. (2003). "Outcome Measures for Clinical Rehabilitation Trials: Impairment, Function, Quality of Life, or Value." *American Journal of Physical Medicine & Rehabilitation*. 82(10 Suppl): S26–31.

Wineberg, H. and Werth, J. L. Jr. (2003). "Physician-Assisted Suicide in Oregon: What Are the Key Factors?" *Death Studies*. 27(6): 501–18.

2

Is Enjoying Life a Good Thing? Quality-of-Life Questions for Jewish Normative Discourse

Noam J. Zohar

In discussions about "quality of life," it is generally assumed that this is a good thing—that (other things being equal) there is positive value in improving people's quality of life. The problems are thought to lie beyond that basic assumption. They are located in knotty questions of definition ("Just what counts as better quality?"), of justice and limits ("How much effort and resources should be invested in order to improve the quality of one individual's life—or that of many?"), and so on. Here, however, I will begin by questioning the initial assumption itself.

Since I am posing this as a question of value, we should distinguish between two very different perspectives—the self-regarding and the other-regarding. I turn, then, first, to the self-regarding perspective.

IMPROVING QUALITY OF LIFE: IS IT A GOOD THING?

Is It Good to Improve One's Own Quality of Life?

When the question is posed this way, it is easiest to see what is meant by "a question of value," as distinct from a question of fact. People *in fact* normally seek to improve their own "quality of life,"[1] but what *value* should be attached to this—what value has been attached to this in the Jewish tradition?

That is to say, to what extent is it seen as a legitimate or worthwhile goal; and how important is it?

Note that it is scarcely plausible, in Jewish traditional terms, to pose a similar question about the value of *life itself*. In the framework of *halakhah*—the Jewish tradition of normative discourse—value is expressed primarily as *mitzvah*, "commandment." Thus preserving one's life is a positive duty,[2] and committing suicide—save in exceptional circumstances—is deemed a grievous sin.[3] But if I am forbidden from taking my life, and am even required to take measures to preserve it, does it follow that it is also a *mitzvah* for me to enjoy it?

In other words: although staying alive is conceived not merely as a matter of self-interest, but rather as having normative value, this need not extend to the quality of one's life. In fact, partaking of life's enjoyments could even be frowned upon, with positive value attached to ascetic self-denial.

Interestingly, the classical talmudic discussion regarding "self-harm" (BT *Bava Kama* 91a–b) includes both of these issues under a single conceptual frame, yet suggests that they might call for divergent judgments. According to the Talmud, there is a long-standing debate: some hold that "a person may not harm himself," whereas others hold that "a person may harm himself." The logical form of the discussion leaves room for a fully autonomous stance, that would allow even suicide or, less radically, allow an individual's choice to suffer permanent bodily injury. But even if these are disallowed, there is a viable position that allows self-degradation (undertaken for financial gain) and self-immolation, driven by religious or moral motives.

Now if self-immolation is permitted, then choosing instead to enjoy life is just that—a personal preference, not a *mitzvah*. A person may choose to promote his or her quality of life rather than to just keep going "as is," but this is no more than the pursuit of (legitimate) self-interest. The talmudic discussion revolves around the biblical model of the Nazirite: an individual who takes a vow not to drink any wine.[4] On the plain sense of the biblical text, such a vow induces holiness. Accordingly, Rabbi Elazar argues:

> One who engages in fasting is called "holy." For if [the Nazirite] who deprives himself[5] of wine alone is called "holy," how much more so one who deprives himself of everything!

Shmuel, however, asserts that "one who engages in fasting is called a 'sinner'," and the Talmud links this to an alternative reading, one which relies rather creatively on the oblique mention of "sinning" in the biblical law of the Nazirite:

> We may reason a-fortiori: For if [the Nazirite] who deprives himself of wine alone is called "sinner," how much more so one who deprives himself of everything![6]

Branding self-deprivation as sinful implies the basic goodness of seeking satisfaction. Aspiring to a better quality of life is laudable, not just egoistic. Against this, the first view praises asceticism and would likely regard any emphasis on "quality of life" as unworthy.

Thus the Talmud presents diametrically opposed views about the value of enjoying life, and conversely of eschewing such enjoyment through ascetic practices. Similarly, it records a variety of views about the value of suffering, including a compelling discussion of "suffering [as a sign] of [divine] love."[7] These debates were continued among medieval rabbis. Rabbi Meir Abulafya (known as "Ramah," Spain, early thirteenth century) endorsed the view that "a person may harm himself," and ascetic practices were by no means a rare feature of Jewish pietistic life. Even for an individual who stops short of active self-immolation, the religiously appropriate reaction to the onset of illness might thus emphatically exclude a focus on preserving "quality of life." Although life itself must be preserved, suffering should be willingly accepted. (Although even when the Talmud speaks of the great rewards ensured to a person who experiences "suffering of [divine] love," Rabbi Yohanan remarks: "[Better to have] neither the suffering nor its rewards!")

Many leading medieval teachers, however, embraced the opposing view: that "a person may not harm himself." Among these, probably most notable is Maimonides, who famously adopted Aristotle's ideal of "the golden mean."[8] Yet it is worth emphasizing that while Maimonides rejects asceticism, he insists that pleasures be sought in moderation, and that the thriving of the body be strictly subservient to spiritual pursuits.

This leads us directly to the subject of the "Evil Inclination," often identified with bodily appetites and egoistic desire. For example, according to one midrashic statement, the evil inclination is older than the good inclination by thirteen years (Eccles. Rabbah 4:13; this number is for males—presumably it is twelve years for females). The evil inclination is there from birth. The picture is very different from that of the "innocent child"; instead, it is that of the greedy baby with a strong and unbridled libido who wants to suck, to eat, to just have fun. Then at the age of discernment, at the age of *bar mitzva* or *bat mitzva*, the good inclination emerges—when it faces an uphill struggle against the longer-established, egoistic "evil inclination." Here our basic bodily functions and simple desires—the very things that figure centrally in any medical promotion of "quality of life"—are seemingly portrayed as something to be transcended in our progressing from "evil" to "good."

This view of the two inclinations is, however, too simplistic. Even if, when left unchanneled, our primary desires are destructive, they are not inherently inimical to life's nobler pursuits. Indeed, the Mishnah (*Berakhot* 9:5) explains that the famous verse, "You shall love the LORD, your God, with all your heart" (Deut. 6:5) means: "with both your inclinations, the good inclination and the evil inclination." Thus the central religious ideal—nourishing our love of God—not only may include the evil inclination, but actually requires its inclusion. Proponents of asceticism might explain this in terms of a challenge: love of God is sustained through overcoming our base inclinations. But the plain sense of this *mishnah* seems closer to the anti-ascetic Jewish tradition. This, then, is a teaching about sublimation: our appetites and primary desires should not be denied or suppressed, but rather should be cultivated and integrated into a mature, loving self. Love of the Divine requires investment of the full human person.

A crude, hedonistic quest for "quality of life" seems incompatible with Jewish traditions. But significant traditional voices would clearly endorse a morally and spiritually mature quest for a good life, integrating physical functions and enjoyments within a spiritual whole. In this sense, there is great positive value in seeking to improve one's quality of life.

That is not to say, however, that any appetite or expensive taste should be fulfilled. In particular, there is a venerable Jewish critique of pursuing *motarot* (literally "extras"), reasonably translated as "luxuries."[9] Critics point out that such a pursuit expresses—and reinforces—character defects; moreover, attaining luxuries comes at the expense of meeting real needs, whether one's own or those of others. I see no reason this should not apply to medical luxuries.

Finally, efforts to improve quality of life must be weighed in the context of the traditional ideal of a life devoted to Torah study. As formulated by Maimonides:

> Every Israelite is under an obligation to study Torah, whether he is poor or rich, in sound health or ailing, in the vigor of youth or very old and feeble. [. . . Each is] under an obligation to set aside a definite period [both] during the day and at night for the study of the Torah. (MT, Laws of Torah Study 1:8)

Now that is the *minimum*; above and beyond it, there is the ideal of a life of scholarship and contemplation:

> He whose heart prompts him to fulfill this duty properly, and to be crowned with the crown of Torah, must not allow his mind to be diverted to other objects. He must not aim at acquiring Torah as well as riches and honor at the same time. (MT, Laws of Torah Study 3:6)

How does such an aspiration stand with respect to a quest for "quality of life"? One kind of answer is to question this very notion of "quality." Surely, to *improve the quality* of one's life must mean to make it *better*, so the issue then is, what makes a life good? Intensive investment in one's physical well-being may thus be seen as contrary to the good life, not on account of an ascetic ideal but because there are better things to do. So, if "quality of life" is understood narrowly, as a measure of the outcome of medical interventions, it may be in competition with spiritual or intellectual ideals, such as Torah study.[10] More broadly, however, such ideals may define what counts as true "quality of life," conferring value on various efforts—including medical interventions—according to their contribution to the goodness of the subject's life.

So we may summarize this first part of our exploration as follows. As with numerous other issues, there is no one uniform approach within the Jewish tradition toward the quest for a better quality of life. Certainly, there is no strong imperative to *improve one's quality* of life, comparable to the imperative of *preserving* life itself. Despite the common perception that Judaism teaches us to celebrate life and the body's enjoyments, the tradition contains a respected view that "a person may harm himself." Moreover, the Jewish ascetic tradition stakes out a path to holiness that might sometimes pass through self-harm, and regularly through self-denial. From this perspective, a focus on one's quality of life can appear as misguided and even degrading.

Against this it seems fair to say that the opposite teaching, that "a person may not harm himself," is the mainstream position, at least in contemporary Jewish communities and scholarship. Pursuit of luxuries—especially expensive luxuries, when others lack basic necessities—should be discouraged, but self-enjoyment is permitted. Indeed, according to prominent Hasidic teachings, a vital aspect of true piety is "worship through worldliness" [*avodah be-gashmiyut*]. But a hedonistic conception of "quality of life" is incompatible with a life of spiritual/intellectual devotion, as represented in the ideal of Torah study.

Does One Have a Duty to Promote Other People's Quality of Life?

It might seem that the answer to this question derives simply and directly from our discussion in the previous section. If it is a good thing for you to pursue a better quality of life, then I ought to help you in that if I reasonably can—that is, without incurring excessive costs or burdens. Likewise, if it is a bad thing for you, then it is wrong for me to assist you in it. And finally, if it is value-neutral for you—neither good nor bad, but simply permitted—then I am also permitted to render help.

In fact, however, things are not so simple. First of all, our value judgments might diverge. To put this in terms of the Judaic tradition, suppose I adhere to the school of thought that values asceticism, whereas you hold that self-denial is sinful. Is it wrong for me to help you improve your quality of life—a pursuit that is worthy by your lights, but unworthy by mine? This is an instance of the challenges entailed by the Rabbinic doctrine of *Elu va-Elu*, "These and those are the living words of God."[11] This issue calls for treatment on a more general level, and I shall not try to explore it here[12]—though it is worth emphasizing that in the healthcare setting, when people argue about quality of life, differences in values too often remain unarticulated or unaddressed.

Moreover, the third of these putative derivations is misguided in the first place. It is not the case that if something is merely permitted to you, then I too am merely permitted—but not obligated—to help you attain it. The fundamental obligation of *tzedakah*—rendering help to those in need—involves meeting others' *needs*, not necessarily or even primarily enabling them to fulfill *obligations*. Of course, it is presumed that these are legitimate needs, but all that means is precisely that they must be permissible—we should not be helping people pursue evil, forbidden things. *Tzedakah* does not require me to support someone's use of crack. But it does require me to help a hungry person get decent food, and it would be absurd to make this conditional upon defining his eating as a duty.

By the same token, if a person seeks medical treatment to improve his or her quality of life, it would be entirely misguided—from a Jewish perspective—for a provider to hesitate on the grounds that "there is no *mitzvah* incumbent upon you to improve your quality of life." The fact that seeking life-saving medical treatment is indeed perceived as a *mitzvah*—and its provision, as a high-priority duty—does not imply that appropriate medical services are restricted to such treatment. The caregiver may be obligated, in his or her professional role, to provide treatment that simply improves quality of life; and helping pay for its provision may well be required, in terms of *tzedakah* owed by individuals or by society as a whole.

QUALITY OF LIFE AS COMPARED TO LIFE PRESERVATION

We have thus established that seeking to improve one's quality of life can be deemed permissible and perhaps even laudable (though this judgment may not be shared by the Jewish ascetic tradition), and that seeking the same for others is even to be deemed a *mitzvah*. Promoting quality of life is, then, a valuable pursuit; but just how valuable is it? One way of answering this question is by comparing that value to the value accorded, in the Jewish tradition, to preventing death—to "life-saving," *pikuah nefesh*.

Preserving life—whether one's own or that of others—at least when in clear and present danger, comes under the heading of "life-saving" and takes precedence over almost all other values and normative considerations. This is classically stated with relation to desecrating *shabbat*—and is meant to apply a fortiori to virtually all prohibitions, equal or lesser in severity. Accordingly, when a medical procedure involves desecrating *shabbat*, the prohibition is immediately and without question set aside, provided that the situation is defined as *pikuah nefesh*, that is, there is (even a slight) risk of a life being lost.

But what if there is no danger to life, but "merely" the danger of losing some bodily part or function? Here traditional teachings are in fact far from clear-cut. If we move down the scale and consider medical procedures aimed just at restoring comfort or removal of pain, it should be recognized that they have definitely not been seen, on the whole, as warranting *shabbat* desecration. And remember—*shabbat* is just the classical example; in principle, the same applies to any other Torah prohibition.[13]

The lesser value attached, in traditional formal *halakhah*, to quality of life per se, is thus expressed in the sixteenth-century classical code, Rabbi Joseph Caro's *Shulhan Arukh*. The relevant section (OH 328) contains two clauses: Clause 9 addresses cases where there is a "way out," following talmudic precedent—danger to a limb can often be defined as a threat to life itself. Clause 17 addresses cases where no such construal is plausible.

> (9) If a person has an ailment in his eyes or [even] in one eye, and there is fluid, or tears flowing because of the pain, or blood flowing etc. . . . *shabbat* should be desecrated for him. [Based on BT, *Avodah Zarah* 28b: "An inflamed eye may be treated on *shabbat*, since eyesight is linked to the power of the heart."]
> (17): If a person falls ill without danger to life . . . *shabbat* may not be desecrated on his behalf in a prohibition *de-orayta*, even if he is at risk of losing an organ.

Is there room, in halakhic discourse, to give greater weight to quality of life in itself? An illuminating argument has been put forward by the nineteenth-century halakhist, Rabbi Shelomo Kluger (also known as Maharshak, 1785–1869):

> I was asked with regard to a person who suffered from the eye ailment called "Schwarzer Star" [amaurosis = partial or total loss of sight without pathology of the eye; caused by disease of optic nerve or retina or brain (N.Z.)], may God spare us, and all the doctors despaired of finding a cure. However in one gentile town they propose to receive him into the gentile hospital, to provide him with healing so that his eyesight not be destroyed entirely—at this stage he is still capable of moving around on his own. He would be forced, however, to eat their non-Kosher food. What is the law—may he do so or not?

First, Kluger admits that the law seems to plainly preclude overriding a Torah prohibition in order to save the patient's eyesight, since the case at hand cannot be construed as involving a risk of death. Nevertheless, he is able to produce an argument based on one of the several reasons offered by the Talmud to justify its basic ruling, that *shabbat* observance must be set aside for life-saving: "It is better to desecrate one *shabbat* on his behalf, so that he will [be able to] observe many *shabbat* days" (BT *Yoma* 85b). Kluger thus reasons:

> Since his blindness will prevent him from the study of Torah and from the [ritual] Torah reading—and the sages have stated that "Study is greater than practice"—therefore, if for the sake of practice it is better to desecrate one

> *shabbat* on his behalf, so that he will [be able to] observe many *shabbat* days, how much more so for the sake of study! It follows that in any case where the [loss of the] endangered limb will prevent the study of Torah, it is certainly permitted to "desecrate one *shabbat* so that he will [be able to] observe many *shabbat* days." This is particularly true regarding blindness, which will certainly prevent [him] from the study of Torah and will also prevent [him from observing] several [other] commandments. (*Hokhmat Shelomo* on SA OH 328)

It is worth noting that Kluger hesitated to rely on this novel reasoning as a sufficient basis for his actual ruling. Still, these lines are an instructive record of a prominent halakhist's struggle to furnish formal grounding for the value of "quality of life." Through the value of the *mitzvot*, he seeks to endow the preservation of bodily function with the value of the religious practices that will remain possible.

For my part, I would propose another line of reasoning, aimed at giving formal recognition to preserving and promoting quality of life. As I emphasized toward the end of the previous section, caring for *another's* quality of life is a *mitzvah*—formally, it comes under Positive Commandment #206. "Love your fellow as yourself" (Lev. 19:18).[14] According to Rabbi Akiva, this is the foundational *mitzvah* of the entire Torah. Hence, even if providing medical treatment aimed at quality-of-life goals does not have the same supreme priority as treatment aimed at life-saving, it does not lack imperative force. If a conflict arises between this duty and another halakhic norm, its proper resolution will not necessarily be self-evident. The *halakhah*—as might be expected of a millennia-old normative tradition—contains second-order rules to provide guidance where first-order norms conflict. Since we are dealing here with a positive commandment, it is appropriate to mention the talmudic dictum that "a positive commandment overrides a negative commandment" (BT *Yevamot* 3b).[15]

Let me make it clear that I am not advocating that our thinking about medical treatment be reduced to application of such formal definitions and rules. Rather, I mean to give expression—in the traditional halakhic medium—to the value of improving patients' quality of life, while recognizing the higher priority of saving patients' lives.

This leads to the final part of the present exploration: direct conflict between maintaining life itself and retaining its quality.

DIRECT CONFLICT: SUFFERING AT LIFE'S END

The great value placed upon preservation of life has an absolutist and hence egalitarian character. In a crucial sense, the same supreme value is ascribed to each and every human life. Well-known examples are the newborn (or even just mostly born) infant on the one hand, and the flickering life of a dying patient on the other hand. There is great beauty and moral force in this absolutism, which is grounded in the notion that humans are created in God's image (what I've elsewhere called "Religious Humanism"; see Zohar 1997, 91–97).

When appreciating this egalitarian humanism, we must admit that its beauty has often been seriously marred by tendencies toward restrictive application. Thus, in various contexts and periods, the life of a "heathen" has been regarded as less valuable as compared to that of an Israelite,[16] or the lives of women as compared to those of men. Against these indecent tendencies, there is a venerable record of internal criticism, insisting on a universal egalitarianism.[17] One can certainly hope that the latter will be persistently endorsed by all participants in the discourse of Jewish bioethics.

In any event, this very same absolutism also often translates into a "sanctity-of-life" approach that rejects any distinctions based on quality of life. On this view, sometimes dubbed "vitalism," every single moment of (human) life is deemed to have infinite value, and life must be preserved and prolonged at any cost.

Now, a broad formulation such as "at any cost" entails more than one problematic implication. Notoriously, it seems to imply limitless spending, with "cost" understood in terms of human and material resources—and if these are committed, without reservation, to saving an individual life, the *cost* will normally be that of *other lives*, whether known or unknown (see Glover, 1977, 210–13). This aspect of the vitalist view is not part of our subject here.[18] Another implication of vitalism is, however, directly related to the present discussion; namely, the position that a person's continued life is more important than his or her own wishes, including a wish to avert great suffering.

In a theistic context, this "sanctity-of-life" view is often cast in terms of divine "ownership" or sovereignty. A person's life, it is asserted, is not his or hers to dispose of according to personal preference and autonomous choice. Rabbi Yehiel Epstein (nineteenth century, Russia) writes:

> Although we see [a patient] suffering greatly in dying so that death is good for him, nevertheless we are forbidden to do anything to hasten his death, as the universe and all within it belongs to the Holy One, Blessed be He, and such is His Exalted will. (*Arukh Ha-Shulhan* 339)

In the second part of the twentieth century, Rabbi Eliezer Waldenberg emphatically extended this to include a case where the patient himself requests an early exit:

> No creature in the world owns a person's soul; this includes also that person himself, who has no license at all regarding his own soul nor is it his property. Thus his granting license can be of no avail concerning something which does not belong to him, but rather it is the property of God, Who alone bestows it and takes it away.[19]

On this view, then, whatever value may attach to improving a patient's quality of life—or specifically, to alleviating suffering—cannot overcome the value of life itself, defined here as God's property.

I have argued elsewhere that, despite the prevalence of this view among contemporary Jewish authors, it represents a departure from mainstream, classical halakhic teachings.[20] True, self-determination does not reign supreme. Suicide is prohibited, since destroying a human being—made in God's image—constitutes a kind of sacrilege. But the same source that formulates the prohibition also defines two exceptions: martyrdom, and a case "like King Saul" (*Genesis Rabbah* 34:9). In martyrdom, a person gives up his or her life to "sanctify God's name."[21] King Saul, according to the biblical account (1 Samuel 31:1–6), fell on his sword so as to avert being tortured to death by his would-be captors.

The sublime value—the sanctity—of human life indeed implied, for the rabbinic sages, that an individual may not take his or her own life. But the prohibition is not absolute, as there are two kinds of exceptions. First, there are situations—strictly delineated—where fidelity to God requires that life be forfeited. And second, there are situations at life's end where the person's own interest in averting suffering is deemed more important than preserving the divine representation inherent in this human life.

Interestingly, some medieval and early modern commentators reinterpreted the "King Saul" paradigm in a manner that essentially conflates it with the "martyrdom" exception: Saul killed himself because he feared that, being tortured, he would commit blasphemy. Most likely, this interpretation was inspired more by the harsh experiences of Jewish communities during the Crusades than by the plain meaning of the biblical text. Still, it is a viable and instructive position, which in effect disallows human self-interest—even in the most dire of circumstances—as a ground for shortening human life.

Other authors, however, retained the "King Saul" exception in its straightforward sense. Nahmanides (thirteenth century, Spain) writes:

> And likewise [inculpable is] an adult who kills himself because of a menace, like King Saul, who killed himself; indeed this was permitted to him. For thus we read in *Genesis Rabbah*: "Could this apply to one who is trapped like Saul?"[22]

On this view, avoiding great suffering at life's end justifies actively shortening life. If we move, instead, to consider refraining from life-prolonging interventions, there may be even more room to take into account the poor quality of whatever extra duration of life might be attained.

My purpose, in this section, is not to provide any detailed guidance on end-of-life decisions. It is merely to show that, even in direct conflict with the supreme value of life-preservation, considerations pertaining to quality of life—at least in the sense of avoiding great suffering—need not always be outweighed. Together with what we have seen in the previous sections, I trust I have established that in Jewish bioethics we should be addressing not only the imperative of life-saving, but also the value of improving quality of life, for ourselves and especially for others.

REFERENCES

Glover, Jonathan (1977), *Causing Death and Saving Lives*, New York: Penguin Books.

Goldstein, Sidney (1989), *Suicide in Rabbinic Literature*, Hoboken, NJ: Ktav.

Goodman, Lenn E. (1976), *Rambam: Readings in the Philosophy of Moses Maimonides*, New York: Viking Press.

Hartman, David (1985), *A Living Covenant: The Innovative Spirit in Traditional Judaism*, New York: The Free Press.

Katz, Jacob (1961), *Exclusiveness and Tolerance: Jewish-Gentile Relations in Medieval and Modern Times*, New York: Schocken Books.

Rackman, Emmanuel (1986), "Priorities in the Right to Life," in *Tradition and Transition*, ed. J. Sacks, London: Jews College, 235–44.

Sagi, Avi (1996), *Elu ve-Elu: A Study on the Meaning of* Halakhic *Discourse* (Hebrew), Tel Aviv: Hakibbutz Hameuchad.

Silverstein, Shraga (1993), translator. *Sefer Hamitzvoth* [*Maimonides*] (2 vols.), New York: Moznaim.

Urbach, Ephraim E. (1961), "Asceticism and Suffering in the Teachings of the Sages," in S. Ettinger, et al., eds., *Yitzhak Baer Jubilee Volume*, Jerusalem, 48–68 (in Hebrew).

——— (1974), *The Sages: Their Concepts and Beliefs*; translated from the Hebrew by Israel Abrahams, Cambridge, MA: Harvard University Press.

Walzer, Michael, et al. (2000), *The Jewish Political Tradition, Vol. 1: Authority*, New Haven: Yale University Press.

——— (2003), *The Jewish Political Tradition, Vol. 2: Membership*, New Haven: Yale University Press.

Zohar, Noam (1997), *Alternatives in Jewish Bioethics*, Albany: SUNY Press.

——— (1998), "Jewish Deliberations on Suicide: Exceptions, Toleration and Assistance," in M. P. Battin, R. Rhodes, & A. Silver, eds., *Physician Assisted Suicide: Expanding the Debate*, New York: Routledge, 362–72.

——— (2003), "Cooperation Despite Disagreement: From Politics to Healthcare," *Bioethics* 17(2), 121–41.

NOTES

1. This does not mean, of course, that people always know how to achieve this, or even that they have a clear realization of what exactly they are seeking (cf. the chapter by Paul Wolpe in this volume). Exploring the nature of "the good life" has been a leitmotif of philosophical quest at least since Socrates.

2. Maimonides, MT, Laws of Murder and Self-Preservation 11:5.

3. The classical rabbinic source prohibiting suicide (and describing exceptions) is *Genesis Rabbah* 34:9. For an extensive discussion of rabbinic teachings on suicide, see Goldstein (1989); and cf. also Zohar (1997, 35–68), (1998).

4. Actually, the Nazirite also vows to avoid corpse-impurity (thus taking upon himself a restriction obligatory for priests) and to let his hair grow uncut; see Numbers 6:1–21.

5. Literally: "causes himself suffering on account of."

6. This second view is cited in BT *Bava Kama* 91b; both views are found in the more elaborate discussion in BT *Ta'anit* 11a.

7. BT *Berakhot* 5a–b; for an illuminating discussion, see Hartman (1985), chapter 8: "Rabbinic Responses to Suffering," 183–203. For a broad historical exposition, see Urbach (1961).

8. Abulafya's position is recorded by Jacob ben Asher in his *Arba'ah Turim*, HM 420. For Maimonides, see MT, Laws of Character Traits chapters 1–4, and particularly 3:1; and the extensive translation from his "Eight Chapters" (accompanied by instructive analysis) in Goodman (1976), 216–61.

9. See, for example, Maimonides, *Guide of the Perplexed* 3:12; and more recently, R. Pinchas Zvichi, *Responsa Ateret Paz* Part 1, Vol. 3, HM 16.

10. This same tension will obtain even if one combines (or even replaces) Torah study with other religious ideals. On the rival traditional views regarding the balance between study and religious praxis, cf. Urbach (1974), 603–20.

11. These words, signifying a fundamentally pluralistic understanding of Torah, are the ACJB motto. For a broad discussion of this idea and its ramifications see Sagi (1996), and Walzer, et al. (2000), chapter 7.

12. See Zohar (2003); for a particular application to a Judaic context, see Zohar (1998).

13. This must be qualified, since halakhic jurisprudence draws a crucial distinction between two classes of commandments: the primary laws of the Torah (= God's Instruction), called *de-orayta*, versus the secondary legislation by the Rabbis, called *de-rabbanan*. It is only the former kind of prohibition that may not be transgressed to alleviate pain. In practice as far as *shabbat* is concerned, most necessary actions can—with some ingenuity—be performed in a manner that excludes them from the formal category of a *de-orayta* prohibition. With regard to other Torah prohibitions, however, this is not easily achieved. The prohibition of desecrating the "dignity of the dead," for example, has been taken by prominent halakhic authors as precluding post-mortem examinations even if these may promise advances in medical knowledge, insofar as such advances are not defined as "life-saving." See Zohar (1997), chapter 5, 123–41. Moreover, addressing hypothetical conflicts-of-values serves to define the relative weight of various considerations.

14. The reference is to the list compiled by Maimonides in his *Sefer Ha-Mitzvot* [Book of the Commandments]. The book was composed in Arabic; an English translation was published by Silverstein (1993).

15. According to the Talmud, this only applies to standard negative commandments, not to those marked for special severity by the biblical sanction of "*karet*" [being "cut off"]. With respect to the examples discussed above by Kluger, desecrating

shabbat is of this more severe category, whereas eating (most) prohibited foods is not.

16. For an exposition of traditional statements and critiques on this issue, see Walzer, et al. (2003), chapter 16: "Gentiles," 441–561.

17. Regarding non-Jews, the classical statements here are by the thirteenth-century Rabbi Menachem Ha-Meiri; see Katz (1961), 114–28. Regarding women, see Rackman (1986).

18. See Zohar (1997), chapter 6, 143–52.

19. This is an excerpt from section 29 of the tractate "*Ramat Rachel*," published in 1965 in vol. 8 of Waldenberg's responsa, *Tsits Eliezer*.

20. See Zohar (1997), 54–58, where I also cite contemporary rabbis who apply the Saul precedent to situations of terminal illness.

21. For a broad array of sources regarding martyrdom in the Jewish tradition, see Walzer et al. (2003), 82–107.

22. See Nahmanides' "*Torat ha-Adam: Sha'ar ha-Hesped*," (Hebrew, *Kitvey ha-Ramban* [= The works of Nahmanides] edited by B. Chavel, Jerusalem: Mosad Harav Kook, 1963–1964, 84); cited also in R. Asher ben Yehiel's rulings on tractate *Mo'ed Katan*, chapter 3, section 94. The same position is recorded by R. Joseph Caro in his *Bet Yosef* annotations to *Yoreh Deah* 345, s.v. "*ve-khen gadol.*"

3

The *Shomer*—Qualities of Mercy and Quality of Life: Arguments from Jewish Bioethics

Laurie Zoloth

INTRODUCTION: THE COMPROMISE OF MODERNITY

The theory of medicine rests on a fragile compromise. On the one hand, the medical gesture of one to another is defined by vulnerability and need. The location of illness is the embodied and particular self, and the perception of illness is both subjective and absolute. Yet on the other, the medical gesture is the sum of a thousand years of inquiry, tradition, authority, scientific method, and discernment, the power to heal based on a premise of objectivity, rationality, and causal outcomes that must be precisely quantified. Thus, the contract is made in modernity—as the power and authority of medicine grows, fueled in large part by the research imperative and clinical trials on human subjects, there has been a corresponding enlargement of the power and authority of the patient to resist and refuse, fueled in large part by the imperatives of bioethics, historical veracities, and the law (*The Belmont Report*, 1979).

It is autonomy—and the practical application of this principle that is enacted in the right of patients to refuse or consent to any medical touch in the clinical context after the full disclosure of risks, benefits, and alternatives—that is the central argument of American clinical ethics. It is how we speak of the moral subject in medicine—as a refusing or consenting actor.[1] Here is how this core narrative classically "reads": The always vulnerable (but rather wise) patient is protected from harm, abuse, and misconduct of the physician (whom we clearly have our doubts about) and the State (whose power is seen as worrisome) which supports, trains, and in many cases manages the care, only by the fragile but unyielding right of the subject to say "no" to even the most powerful.[2] All of medicine's power is held at a protective distance by this "no," the possibility of negation, the negative right.[3]

This right has been first formulated on behalf of the adult, rational patient, and with the use of proxy mechanisms,[4] finally to formerly competent persons who are bound to follow the wishes directed in advance.[5] So powerful is the assertion that the right to refuse must be honored (Dresser, 1990, 425), that parents may assert it on behalf of children, acting in their best interests (see Committee on Bioethics, 1995). Bioethicists crafted this response to a world horrified by the doctors of Nuremberg, who violated the bodies of their victims under the guise of "medical experimentation," beginning with the elimination of the disabled. It is in response to the casual abuse of life's sacred dignity in the Shoah that bioethics opposed the linking of this right to people with "good" qualities of life, or cognitive ability (see Freedman, 1999, 146–49). We remember that the German doctors turned away from patients—first by declaring their lives to be useless, "dis-qualified" by mental disability, next, by experimentation on the bodies of the voiceless and vulnerable, and we honor that memory with regulatory protection and with a promise to fiercely protect patients who are vulnerable (see Freedman, 1999, 146–49).

The conference from which this volume is drawn asked a specific—and a very American—question: How should "quality of life" influence our end of life treatment decisions? In a bioethics so focused on autonomy and the "right to die" what is the moral force of such a question? How does an American *Jewish* bioethics reflect on such a right? (And, is this even a Jewish question?)

Where are the coordinates and dissimilarities of law and ethics? How is the moral subject negotiated by *halakhah*? And finally, what can we learn from the methodological frame of Jewish ethics? Let me argue that it can be said to be the case that both premise and method will be core distinctions, but let me first show you how: Because thinking about specific cases is one of the methodological moves I want to convince you is key, let me turn to three cases where treatment refusal is a question and in which quality of life is an issue. We will be particularly concerned with cognition (we are Jews, after all, and memory and language are central to how we regard the moral subject) and the question of treatment refusal when the loss of cognitive ability is at stake. Let us turn to some specific cases—stories—to think, perhaps midrashically, on our topic:

Meeting Robert Wendland

It was the earliest of spring in California, meaning that it was February, and the almond trees that crowd the great Central Valley, row on row, had filled the sky with low, lacy white haze clouds. I rolled the window down and breathed in the unbearably sweet air, still cool in the early morning. My colleague and I had been driving along the arrow-straight highway for hours; we were off to Lodi, a tiny farming city off the Sacramento River delta, to meet Robert Wendland, or rather to meet his family. Mr. Wendland lay waiting for us in a nursing home bed, still under the pink cover. His wife met us outside, and took us into the room. We were there because we were figuring out whether we would be pro-bono experts in what would become the Wendland Case, another in a series of cases in which the public tragedy of significant brain damage after a terrible accident led to an agonizing public trial to decide a personal dilemma: whether a person would be better off dead than living in a persistent vegetative state.

But before he was a public case, an event at which people picketed, and bioethicists were interviewed on television, Robert Wendland had been a big, active man, a truck driver for a parts company, with two teenage daughters and a pre-teen son. There were a few distinctive things about Mr. Wendland: he had overcome an abusive childhood, he was a recovering alcoholic who recovered only partially, he was close to his brother, he had been married a long time for a Californian, and he recently had been exploring a return to Roman Catholicism. He had been clear at the moment of most trauma in his life—when he and his wife had decided to discontinue life support for his own father-in-law a month before his own accident—that as a Roman Catholic, he believed in the afterlife, and he believed in the Church's teaching that extraordinary measures were not always required if they were burdensome and not commensurate with improvements in a good "quality of life" according to his wife. It was a profoundly Roman Catholic view honed by the case of Karen Ann Quinlan. "If I cannot be a husband and dad, I don't want to be here," he said, recalled his wife.

Wendland and Quinlan, Cruzan and Schiavo became events for American bioethics, yet before each of these cases came to the court, each family faced, with their priests and ministers, a complex negotiation about the meaning of maintaining a life that seemed so diminished in quality. In many cases, they disagree—most famously, of course, in the case of Ms. Teresa Schiavo, who suffered traumatic anoxic brain injury in 1990, entered a permanent vegetative state shortly after, and was taken off life support after a fierce court battle in 2005. Her parents saw her life as having infinite value, regardless of quality, her husband came to see her as no longer a person capable of any response. The cases of Robert Wendland, Ms. Schiavo, or other families challenge us in bioethics and challenge us as Jewish ethicists, for our texts did not envision a world of ventilators and feeding tubes, or of permanently unconscious but alive people, let alone people who anticipated such a state and asked not to be left in it. The ethical issue is whether one should consider quality of life when thinking about treatment refusal, and if so, who determines and who assesses such quality?

The Question of Research on Dementia Using Demented Patients

At times the clinical decisions about life's quality are far more subtle: In a second series of cases in which I consulted, an earnest young researcher had been advised by the IRB to ask for help from some ethicists. He wanted to study memory loss in the elderly, and I was asked to reflect on whether elderly, senile patients can be used as participants in research about senility in the elderly. I was struck by this problem—for it was akin to the familiar issues of consent: consciousness, moral status, and surrogate decision making for profoundly ill persons whose quality of life suggested that perhaps they might be vulnerable. Moreover, since what was being suggested was the use of the patient as a subject for research, they were especially vulnerable as targets for research abuse. In this case, what would the subjects have wanted? How could a surrogate respond?

Rules for clinical trials in bioethics, even more directly than in patient care, are based on a potent but covert theory of evil. The concern of bioethics is that the clinical trial must be free of any financial or academic conflicts of interest, cannot be framed by greed or mendacity or negligence, and must at all times be regulated by others with attention to the protection of the vulnerable subject. The regulations that emerge are shaped by the worst history—Nuremberg, in which that was surely the case—the Jewish body was inscribed, tattooed, numbered, and named as research organism by the most prominent researchers and physicians of the period. Bioethics reminds physicians of such things as "battery" when teaching the behaviors of clinical research, so conscious we are of the power of medical intervention, even in these situations of research. The ethical issue is that the only way to do serious research on senility is to do research on senile persons, and whether one can consider the quality of life experienced by these subjects as a part of the decision making process, and if so, who determines and who assesses such quality of life?

The Easy Case Is Not So Easy

The final case raises complex issues in its apparent straightforwardness. The chair of the ethics committee is nearly embarrassed to call, she says—it is such an easy case. And it is—a severely disabled man wants to have life support removed—he is quadriplegic, and hates his "poor quality of life."

She was right, of course. It was an easy case, it was like a hundred you can read about, in which a patient is suddenly snatched from a vivid life into our intensive care unit, it is a fire, it is a heart attack, it is a fever, and this time, it was one jump into one country swimming hole, a flight taken all this man's life, but this one time it would land him, broken at the high neckline, C-4, flat in bed, the breath on that leap his last unassisted one. He wants to die because we have failed him and he cannot imagine a life beyond this one we have given him, and he yearns for his old life back, a thing we call denial. We cannot stop his pain very well, he hates his passivity, he is a quadriplegic, and that is who he has become, his name, "the quad," "the case," and he would rather be dead. The rule of the HMO had been rigid—he did not qualify for a motorized wheelchair, nor for rigorous therapies she had tried to fight for, the pain control they wanted to try was "off formulary" (meaning that it was not used in the HMO pharmacy and that a generic, older, and thus less expensive pain medication was used). It would be difficult to obtain. The ethical question here is whether the quality of life this patient experiences is so poor, of such limited value to him, that he would rather be dead? Here again, who should decide about life's quality? How is this discerned and assessed?

Or is this the correct ethical question in this case—or in any of the cases I noted briefly above? What would it mean to ask a *broader* question—how can medicine itself be transformed, not how bioethics can somehow help patients escape from it. It would be to think about duties, and not rights, and that would return us to beneficence—how is it that we do good? How can we transform the way to perform medicine? Unless we do, what will make our case, the enactment of the party line of bioethics, so disturbing is that we become accomplices in the death itself—by so doing, we endorse the allocation decisions that force the choice—the limits of the hospital formulary, the limits of the insurance policy, the cost of health care aids. We, in this way, play with the house, we are "in-house" bioethicists—which is surely the case in American bioethics, for we are called by, paid by, and supported in the field by our affiliation with the medical school or hospital.

In the cases presented to us,[6] we are asked to consider several problems. Let us address each separately, unpacking the question usually portrayed as the "right to refuse treatment." First is the consideration and the assessment of the quality of life of the patient. Second is the problem of whether such a consideration can be validly made and of what importance one attaches to such a determination. Finally, there is the question of whether and under what circumstances treatment can ever be refused. In the case of experimentation, a further complexity is introduced—may intervention be initiated based on the same arguments that allow it to be refused?

And at these moments, when I have been asked to reflect on what a Jewish argument from our texts of Hebrew scripture, Talmud, responsa literature, and historical or minhagic tradition might add to our understanding of the issue, I struggle to locate myself, as many of you do, on the narrow ridge of faith and reason, of a praxis of Orthodoxy and an allegiance to American bioethics.

Why was it that I would be the pro-bono expert for Robert Wendland, but wondered if I should sign the amicus brief in the Schiavo case?[7] What are our obligations to the incompetent patient and how might these guide our understanding of the question: What is the Jewish view on life support or participation in experimentation, and what are the obligations of the Jewish physician to provide informed consent and to protect the autonomy rights as the trumping ones, given that autonomy and its defense requires, of course, cognition, reason, and directed will? How can perspectives from the Jewish legal tradition influence the discourse on the cases before us, ones in which we are asked whether the measure of life's quality ought to be a moral consideration of our contentions?

Let me argue here that, to a large extent, the concern about life's quality and relational ability, which is such a large concern for our Catholic colleagues and for secular discourse partners, is not the only question that is asked about such cases—what is at stake is not the limits of autonomy, but the scope of the duty to care for one another despite levels of consciousness of the person in need of care, and always, always, the problem of justice. In a sense, the case of consent in the case of medical experimentation, or in life support withdrawal can be considered as the essential problem of consent to any embodied action taken to its most extreme. The incompetent patient is merely the most vulnerable patient in the medical context, for she is unable even to marshal the fragile defense of autonomous contractual ability in her defense. End-of-life care is merely the last form of medical intervention, unable even to offer the compelling notion that nearly all activities undertaken for healing or restoring a life are obligatory for physicians to provide and for patients to accept.

Let me further suggest my answer and then argue my way back to it. There are indeed many other scholars—Brody, Dorff, Tendler, Zohar, Newman, Green, Weisbard, and Freedman—who have found reliable texts that describe how one can lose one's "*ta'am*" (meaning both a literal "taste for life" and a "sense") for life. Brody, in his defense of some limited forms of rational suicide, and Tendler, in his discussion about end-of-life care, cite that narrative of an elderly woman who, tired of existence, asks to die, and is told to stop her daily prayer, and when she does, she dies peacefully. Other texts define different levels of obligation, in some limited settings, for doomed or terminally ill persons. Using these examples, Dorff has made a case for treatment restriction (Dorff, 1998). Further, the narratives of the Law are replete with accounts of aging and death, in which there are no last-ditch struggles using the mendicants of the period. Clearly, the talmudic and midrashic texts are not naive about aging, senility, or the world of loss that faces the disabled, even when average life expectancy was forty.

But since I believe that concerns about life's quality are largely about the loss of cognition and control, and that they are always framed by our struggles with allocation and scarcity, I believe that instead of arguing that if life's quality is too poor or fails to meet a certain standard, we can end it, I want to think about how we think about thinking, consenting, and the being that is created by this sort of "being." In part, this is because I must add that from a Jewish perspective, these cases are not particularly difficult—in fact, I would argue that even amidst the intense contention of the tradition, the halakhic response is fairly straightforward (see Rosner, 1991, 65, 75–77). Since healing is a primary and role-specific obligation held by both the patient and the physician, treatment that offers a prospect of healing must be offered and accepted. The non-rational person cannot make contracts or promise on his own, and others must protect his interests in all such negotiations. Parents must be honored and cared for even if they become senile, and finally, the body is not the possession of any person, not even the self, since the body belongs only to God. Further, if the view of the doctor in secular bioethics is surprisingly negative, the view of the doctor in modern Jewish bioethics has been cheerily optimistic (Jakobovits, 1959, 291–92). Finally, Jewish bioethics suggests we can shift the gaze from the *will* of the patient to the wider world of her family and community, which means that apprehensions of subjective phenomena are ultimately social. In this account, the patient is not the only subject of the gaze of medicine, and certainly not of ethics. She is always a part of a family and a complex social relationship of duty, doubt, and choice.

Since Jewish method is textual and linguistic and case-based, in this paper let me turn my attention to why Jewish ethical norms differ so considerably from those of liberal theories that are the ground for normative bioethics, and let me, in the name of full disclosure, note that the Jewish ethical praxis to which I am personally committed applies in this binding way only within the Jewish community; frankly, I think that the rabbis are right and indeed hope that they can justify and convince a larger polity because of the plain good sense of their direction.

In this paper, I want to undertake an extension of an idea that the late Benjamin Freedman advanced in his seminal work on the topic, *Duty and Healing* (Freedman, 1999), that of the utter centrality of duty, and of the role of the duty of the patient to participate fully in her care regardless of her quality of life. In this, I will note first that Jewish law does indeed have categories of distinction for the patient who is mentally unable to participate in cognitive activity, and hence who falls into a particular category of illness—(in this sense, there are some quality-of-life considerations) but that by far the most important considerations of all treatment negotiations of this type are ones that consider the duty of the guardians—the *shomrim* [= plural of *shomer*]—for care and moral authority.

QUALITY OF COGNITION AND MENTAL IMPAIRMENT IN THE JEWISH TRADITION

The most extensive treatment of how a quality of life determination might indeed be applicable in treatment decisions is in the case of mental incompetence and the particular cases of how mental incompetence might affect decision making. The implications of these texts for Jewish bioethics have been put forward by David Bleich in his consideration of the case of research on mentally incompetent persons (Bleich, 1983). Bleich begins by discussing the criteria for assessing the legal standing of a divorce effected by an irrational or mentally impaired husband, a problem in Jewish law, since the system of contracts, including the marriage contract, requires rational actors, and it is the husband who must competently dissolve the marriage. What should be done when a person loses cognitive ability? He claims that such a case is useful for our purposes in deciding other cases of relationships and duty when one partner is not fully able to make decisions. Linked to his argument is the recognition that losing one's rational faculties would be a tremendous loss of life's quality—not only is rationality a core feature of human life for the Jew, a thoughtful marriage is also critically important. What is at stake in this argument is the ability to sustain a *marital* contract or not—what is the relationship expected when one is dealing with an ill person who cannot enact the commanded acts that are required of him—what is the moral standing of a moral agent with diminished capacity? Such a question is also at stake in how one treats the ill person or *shoteh*—unclear under the covenantal obligations of the Law. What are the *moral* status and the *legal* standing of persons who cannot enact the *mitzvot* (because they are too ill mentally), which are at the core of Jewish religious and social life? But how ill, or of what poor quality must the life of the moral actor be before it is a consideration for the law?

Bleich (1983) cites the Mishnah Terumot 1:3:

> Who is a *shoteh?* One who goes out alone at night; one who spends the night in a cemetery; one who tears his garments.

Evidently, through these examples, the text is seeking to describe and define one who is acting with reckless abandon, risking her life for no apparent benefit, or in both of the last examples, one who has a distorted relationship with death, or perhaps despair, torn garments being a classic sign of mourning in Jewish tradition. But in the text that immediately follows, there is a dispute on several grounds among the other rabbinic interlocutors—are all of these actions necessary for the declaration of incompetence, or only one of them? What of the person who can give logical reasons for self-destructive behavior? In the Gemara that comments on this, BT *Hagigah* 3b, the general argument is advanced that *any* behavior by a person who "destroys all that is given to him" marks the *shoteh*.

For Maimonides, who also considers this text and its use in medical decision making, it is the standing outside of the habitation of the commandment that defines the status of the *shoteh*. This means that if the ill person, or *shoteh*, is a person that cannot enact the commandments, he is too ill to decide about their applications. Even if the person "converses and asks questions to the point in respect to other matters, he is disqualified [from serving as a witness] and is considered to be among the *shotim*."[8] For Joseph Caro, and other commentators, including Rashi, the bizarre behavior that does not allow full compliance with the Law is correlative to other contracts and covenantal relationships "exempt from the commandments and from penalty, whose acquisition is not an acquisition and whose sale is not a sale."[9] Such a person cannot serve as a witness. Yet, some contracts and some activities are permitted to a person with a diminished capacity, and the textual tradition struggles with how to evaluate the level of competence needed for reasonable moral decisions (Rosner, 1979, 387, 394–95). For example, in the cases of the mentally retarded, even if their disability is relatively minor and allows partial compliance with *mitzvot*, any ability to perform the *mitzvot* should be fully encouraged ("if he can speak, his father must teach him Torah and the reading of the *shema*").

But the commentators who reflect on this text are primarily concerned with establishing a category in Jewish law of a person, a *shoteh*, whose quality of cognition, or perhaps, then, one for whom the quality of life, is the determining factor in their obligation.

But what of the persons who then support and surround such a person? For us the responding community, the vulnerable one is like all Others—only more visible—there is no end to our duties to them. We would never allow them to be lost in an interior world apart from the judgment of the rational community. It is not a favor to them, or a freedom to them, to allow them to live in a degraded manner, and this is true for all whose quality of life is diminished—parents, children, and teachers—to whom we owe special duties.

The point of this is that autonomy is a poor substitute for the responsibility of the responding community. We might note that to be judged incompetent in a Jewish sense is to be unable to perform the social role that is needed in the particular social contract that is medical consent and refusal. Other social roles may remain perfectly intact and can be so honored, but the capacity to act as a contractor of services, a rational adult, a responding patient cannot be claimed, and in this, our cases are framed by the limits of Jewish law and definitions (Freedman, "Competence," 1999, 240).

How is one defined as ill in this way—particularly needing our communal support? What the actions listed in the Talmud and in later responsa literature suggest is the utter abandonment for *self-regarding protection*, for full flourishing, and perhaps most importantly for the inability to use rational language to defend one's choices to others to whom one is in a relational community. For bioethicist Benjamin Freedman, the key ability, to give consent or to refuse treatment or experimental participation, is to "play the patient game," using a proviso he named "recognizable reasons" in the decision making process (p. 242).

SPECIAL OBLIGATIONS TO THE PARENT WITH DIMINISHING CAPACITY

Since so many of the cases of end of life and quality of life considerations occur within the parent-child relationship (witness Schiavo, Cruzan, and Quinlan), it is a feature of Jewish law that when the patient is one's own parent, the burden of responsibility is greater (Meier, 1986, 75). One owes parents both honor (*kavod*) and reverence (*mora*). Meier defines honor as positive, tangible acts of personal embodied service and reverence as largely protective—avoiding acts of disrespect or contradiction.[10]

The Gemara that discussed this uses as a basis for discussion the following story.

> Once he, [Dama son of Netinah] was seated among the great men of Rome, dressed in a gold embroidered silk garment, when his mother came and tore the garment from him, slapped him on the head and spat in his face—but he did not shame her. (BT *Kiddushin* 31a)

A long discussion follows—the examples of the insult get progressively more extreme—and the aging parent is depicted as odd, then wild, then abusive, but even here in this text, a parent is not abandoned completely. In one event, recalled in one source and intensely debated, with significant commentators on each side, limits are finally raised to filial toleration, but not to responsibility—if a child finally leaves an abusing elderly parent, he must get caregivers for her. But let us be clear: the view of the family assumes duty, awe, and love—and if they falter in a private way in love, they are caught by the web of the Law (BT *Kiddushin* 31a).

THE LIMITS OF AUTONOMOUS CONSENT AND REFUSAL IN JEWISH BIOETHICS

Interestingly enough, one of the mesmerizing debates for bioethics has been that of informed consent of the patient in just this situation—and whether a diminished quality of life can trigger the end of life support if there is consent. Freedman noted in 1999 that in the over twenty books in English that describe themselves as compilations of Jewish ethics, none has an entry on informed consent. As Freedman says: "Full disclosure of a patient's condition is not required of Jewish doctors to Jewish patients if such a disclosure would destroy their hope for recovery." However, the proper assessment of risk must be understood (see Bleich, 1991, 31, 62). It is the duty of the patient to select a wise physician, and it is traditionally on this basis that the foundation for trust is laid (see Freedman, 1999, 142).

As Freedman says: "[i]n brief; [t]he absence of a rabbinic doctrine of informed consent is predicated on the view that, in general, the patient's illness is diagnosed with certainty, and medical knowledge dictates that illness be treated in a single manner" (p. 174). He cites the literature as follows:

> [T]he entire value foundation underlying the principle of informed consent is almost totally nullified. According to *halakhah*, the mode of treatment is frequently not established according to the will of the patient and his consent, but rather according to the objective situation . . . [t]herefore, regarding an ill person or evident injury, when the doctor has certain knowledge and clear acquaintance, and employs definitive and proven treatment [*refuah b'duka ug'mura*], certainly the ill person who refuses treatment in the case of danger is coerced, and [his express refusal] is not accepted (p. 154).

Informing a patient is described as explaining the intervention that needs to be performed (p. 178).

But in American bioethics, autonomy is all about consent, since individual liberty and not social justice was the one thing codified. The notion of a self who sets rules for the subject in her relationship with the world implies a being with rights, a theory of rights themselves, and a linguistic account of both positive and negative rights (Taylor, 1989, 33–36). Many formative assumptions in analytic philosophy require an autonomous, consenting and refusing subject and a rational and language making self if they are to be valid (Rawls, 1972, 433–39). Finally, all ideas about autonomy rest on the sense that the good can be personally apprehended and enacted entirely by this entitled, knowing subject, for it is only this self that has the power to resist heteronymous authority, only the dignity of this knowing self, who knows, or who is enlightened by, in the classic sense that John Locke (1960, 445–46) and others understood, the good that cannot be abrogated. Consent is the thin reed of protection from a history of others' intrusion (Borowitz, 1992). For Kant, this tension between autonomy and community was resolved by the formulation of the rightly ordered will of the self as being consonant with the universal will that could be held in common by all members of a universal community, a categorical imperative that all would understand as rational and justifiable (pp. 9–10). A key marker of rationality was whether the view would compel assent and allegiance by any listener.

The entitled self is one that is required by modernity, a self who is free prior to any obligations and free to choose among them. One's self is prior even to family, to nation—which is why only fungible contracts can be ultimately arranged, for if the self is primary, then there is always a sort of escape clause. This self must give the consent of the governed to any who would enter the free sphere possessed by the free self. Linked to this idea, and central to consent in medicine is the notion that the body is another sort of possession, and that this ownership is inviolate. The duties such a self might incur are only those duties that one undertakes to maintain one's rights, based on a concomitant theory of rights, and the relationship to the other is in service to this idea. The notion that there is an objective good or truth is queried by this autonomous self. Hence, if I and I centrally (if not utterly) can see and understand the good, then unless I can rationally convince you that we in fact share the good act, there cannot be any but a set of subjective standards of behavior or belief, leaving us with only two choices—toleration, one logical outcome of this Enlightenment standard, or war, rather what George IV had in mind. In the clinical setting, one seems to be faced with a similar binary choice—toleration of what might seem to be a dangerous subjectivity, or the forceful enactment of the external good, the categorical imperative of all but the knowing subject, with the full violence of the state if need be. Rights emerge from an experience of oppression, hence they suggest both an ontic metaphor (the farmer defending his farm) and a view of evil (it is the other who is a danger), and this view informs bioethics as well.

But for Jewish philosophy, the argument for autonomy has been thin, held against a robust and vivid construction of both a heteronymous set of norms (Seeskin, 1992, 21, 33–36) and a community process in which one is obligated to enact and perform these norms (Borowitz, 1992, 9–10). Hence, a series of alternate views about the self as an autonomous being might be useful. First, that it is rational will that leads to an acceptance of the commandments called the *mitzvot*, and that the freedom of the self only exists subsequent to the obligated and listening self. One is responsive in this formulation even prior to the Commanded Law, by virtue of being part of a community defined by its ability to do this very thing. Thus, duties are prior to rights and these duties are not subjectively defined, but are defined ultimately by the contention, debate, and linguistic-moral analysis of the interpretive community. It is not only that the Law is prior to freedom, but that the historically linked respondent and responding community is prior to freedom, and importantly for clinical medicine, that the Law is always linked to issues of justice, to positive obligations that place limits on the freedom to use resources. Duties are not the result of a history of personal experience. The duty exists as a priori—why? Not only because one is divinely commanded, but is commanded in public, as Emmanuel Levinas (1990) notes, in the full view of the one who stands at your side. To be a Jew is to have witnessed the giving and receiving of the duty.

Unlike the Enlightenment thinkers, the body in Jewish Law is not the corporeal possession or expression of the self, but the entire self, mind, body, self, *nefesh v'guf*, is the possession of G-d (see Freedman, 1999, 175–76, 183), to be held in temporally tangential trust for a period yet to be specified (Levinas, 1990, 211, 224–46). It is a central feature of Jewish case law, and a commonplace (Zoloth, 1999, 86–89) in the contemporary Jewish bioethics literature, that this is the case (see, for example, Dorff, 1998 or Freedman, 1999). It is the reason, for example, that one cannot testify, be questioned, or confess about one's own actions in a capital trial, because one cannot forfeit a life which is not one's to give up. One's body, one's land, one's future, the very self of the Jew is not a "transaction" that is completed and then a thing to be owned, to be used in any manner, to be consumed by personal desire. In point of fact, it is this embodied quality of the relationship that allows the infinite possibility of redemptive action—it is not a contract between strangers; it is a covenant in which the obligations are not abrogated if a party fails to deliver the goods. The body itself is a promise of a relationship that is based on obligations, habitual activities, and daily repeated tasks. For Maimonides, and for the rabbinic commentators of the Gemara, the necessity of persisting in moral activities and in prayer is important even when one does not feel inner motivation (see Borowitz, 1990, 23–25). Such a duty toward G-d and toward your fellow person emphasizes the duty over the freedom of moral agency. Nothing can be surrendered, nothing neglected or abandoned, no one not enmeshed in this promise, received in the hearing and witnessing of each, and then made to the other. The enormity of the gift, the largeness of it, life itself, notes Emmanuel Levinas (1990), is only matched by the ceaseless obligation of the endless debt that the ethical system of *mitzvot* suggests. Hence there is no autonomy without the prior need for duty and justice in relationship, which is not some liturgical abstraction, but the demand of personal attention and care (see Freedman, 1999, 175–76, 183).

In other systems of natural law, for example, in Catholic moral theology (see, for example, Kass, 1985), the autonomous self is linked to the idea of Natural Law. Natural rights to oppose the rule of earthly kings are critical if one takes seriously the primacy of the Cross (Hauerwas, 1986, 8–14). Much of the authority of such a natural law rests on the idea that the individual can apprehend, understand, and reason toward the good—to be a person is to be able to reason toward the good (Engelhardt, 1999, 113, 126–27 n2). The idea of a deist endowment of equality and enlightened conscience also undergirds liberal theory from Locke, Rousseau, and Jefferson to constitutionally mediated rights.[11] Such rights theory and consent springs from the right to rebel and resist. It is the drama of privacy, protection of property, and the self understood as a private and hence a protected sphere that allows for this notion.[12] Constitutional guarantees that prohibit the intrusion of the state into this protected and private sphere, and case law that draws the clinical encounter into that sphere, then valorize privacy, and the freedom to resist, law-by-law, as central to the self.

For Jews, what is at issue in how one is valued as a citizen of a community is by Law, covenant, not authority, since submission to heteronymous, and not autonomous norms is assumed. The Jews were bound to the Law and to a system of judgment in the place of pharaonic despotism—at stake is not kingship, but injustice. Hence the just interpretation of the law, the systematic resolution of disputes and the commandments are given simultaneously.[13] It is for this reason that Levinas reminds us of the rabbinic notion—stated by Rabbi Shimon—that the covenant at Sinai is also made in the tent of Moses in the wilderness, as the judges appointed used the Law (Levinas, 1990, Abrams, 1998, 211). Law is not natural or inherent, and dignity is not tied to nature but to faithfulness to the Yoke of the Kingdom of Heaven (see Urbach, 1988, 262–64).

We are social beings, intended for polity and civility, intended for the word, for the open, ambivalent text, hence needing the interpretive community. The Law is a matter of social ethics, in that even private decisions are subject to its scrutiny. Impulses of personal cognition are important only if rational and justifiable to a moral community bound by the Law.[14] For Robert Gibbs, commenting on this aspect of Levinas, this is key.

> If ethics can no longer originate in an autonomous self, then ethical responsibility need not dissolve into a puddle of ideological constraints. Rather, ethics stands in need of a reorientation, one which can loosely be called social ethics; that is, responsibility emerges in an already social interaction. The responsible person is already bound with others in a material nexus (Gibbs, 1992).

This idea is not exclusively Jewish, but it is excessively so.

Maimonides notes that there can be no individual flourishing without community engagement (see Urbach, 1988): moral improvement is attained by practice of paradigmatic actions (Gibbs, 1992). For Maimonides, lawlessness is the social resultant of each person judging his own case (p. 82).

THE SOURCE OF DIGNITY: CONTRACT, LIMITS, AND RESPECT FOR PERSONS

Does it demean a person not to allow him to be fully self-legislating (Abrams, 1998, 66–67)? Not in this Jewish construct—even for the fully rational person, many choices are constrained by role-specific duty. It is clear that one can fully respect persons and yet limit the horizon of possible social relationships and duties. For example, such a limitation applies to the deaf, the one who cannot walk, the one who has impaired hands, and to the blind, all of whom cannot testify in a court of law—yet these are respected as citizens within this limitation on actionable narrative witness. Judith Abrams (1998) points out that even able bodied adult men were once children and thus subject to some limitation. Hence, the source of "dignity" is not utter equality in all activities.

There are other constraints on the very premise of informed consent in advance directives, for the making of vows is itself a problem. This is implied in the Biblical text, when the idea of the Nazirite, the emblem of vow-making itself, is seen as a troubling figure. Rabbinic commentary then takes up the theme that vow-making in general is somehow suspect as well. In this sense, an informed consent decision *in advance* is a sort of a vow, one undertaken on behalf of the person that one is anticipated to become. Hence, if one writes a statement in advance of the actual arrival of the self that would make the decision, and that self is protected in advance by the self that no longer exists—a frail defense in a world of strangers, it can be seen as troubling as well. All of which is to note that the two concepts that frame the entire debate about informed consent, surrogate decision making and dignity at the end of life are very differently understood in Jewish thought. Thus, where do we turn for a more robust sense of Jewish ethics in these cases of treatment withdrawal and research initiation for the person with respect to the "quality of life concerns" queried by this volume? Let me suggest an alternative, and let us turn to it in the second part of this essay.

WE HAVE SET WATCHMEN ON THE WALLS TO GUARD OUR SISTER[15]

Benjamin Freedman anticipates the current debates about treatment withdrawal and quality of life assessments when he writes of the limits of a halakhic regard for consent even in the case of reasonable and efficacious clinical interventions. Freedman suggests that we ought to understand the duty to heal as directing our activity toward the other, but to oneself as well. In thinking, as I have just done, about the primacy of autonomy and the informed consent discussion in American bioethics, Freedman puzzles over why something so utterly central to clinical and research practice could be so utterly absent from Jewish texts. In his work, therefore he seeks a middle ground.

Let me expand his work further to suggest that there is a central philosophical principle, a Jewish idea running throughout the talmudic tradition that lies, I would argue, precisely *between* both sides in the vexing argument in American bioethics over which principle should trump in an ethical dilemma, beneficence or autonomy. The core narrative then, is not, and ought not be autonomy, but stewardship, or what I will call, after Freedman, being a *shomer* [or in the female form, *shomeret*]. Freedman calls this a bailee, or a caretaker, as does Bleich (Freedman, 1999, 175; Bleich, 1998, 65).

It is not only over the body that one has a positive duty to "carefully safeguard"—it is also over one's life and the goals of one's life in general (see Urbach, 1988). One "has" one's life in a particular sense—the having of it is not a gift, since you do not get to keep it, and since not all activity is permitted with it. It is not a possession, since it really belongs to God.[16] It is not a loan, since one does not pay interest, and since the rules for its use are specified, and since you did not in this sense ask for it. It can be understood as being similar to that broader class of entities that one's neighbor leaves with you and asks you to hold for safekeeping.

For this category of "having," there is a complex set of rules: how to store and care for the item, how to be fully clear about the limits of use and of care of the item, and the terms of the reacquisition of the item to the owner. One cannot hold a neighbor's coat and mistreat it or harm it, and one cannot hold a neighbor's animal and mistreat it—it must be returned in the same condition as it was placed in the care of the guardian. One must seek careful advice on how to use the object as well: in monetary matters, one who is given money to guard must seek out the best advice about how to use it (Freedman, 1999). One is also obligated to act as a *shomer*, a guardian for disempowered persons. The relationship is used to describe one's guardianship of the land—and extrapolated to persons.

In American law, property is defined by exclusion: "a bundle of rights intended to exclude others."[17] This is a definition that allows us to understand that intellectual property and manufactured goods can be understood as similar to each other and to land. Note the contrast with the idea of agricultural possession in the Jewish texts: here, as in Genesis, one is a guardian of the earth outside of Gan Eden, and later in Leviticus and Deuteronomy, one is a guardian for the Land of Israel. The Land is held in trust for God, and this holding defines one's obligation to the poor. We are reminded that we hold the land and that we do not "own" it in the act of debt forgiveness and charity. While one's social location to the land will vary depending on fate or chance, the poor are never entirely "without" the land's bounty. In this sense, the advantaged landholder must accept a complex net of securities that transfer the goods of the land to the poor—it is God's land, and God will feed the poor, always.[18] The privileged landowner does not then give the poor his leavings because of pity, but he and the poor, his brother-by-his-side, in the classic construction of the Law, are both participants in the process of caretaking of the gift of the Land. The way that production is structured creates what in a purely marketplace sense would be "waste" (hence left behind) but in this sense is a *different part of the productive harvest,* the part assigned to the landless. Since *all* in the society, even men, must at some point be children, all must at some point be needy ones, dependent ones, needing a just *shomer* to watch for their best interests. This dependency is a part of our ontic nature. In Jewish law, then, the issue is not *whether* to include other people in defining "possession" of the Land, it is only *who* to include in the bundle of duties that property is. The inclusion is very large—nearly all are invited in during the seventh year resting or Sabbatical Year—for example. Here, the Other is not the intruder, the Other is one's neighbor, or one's own brother.

There is a duty to protect one's holdings from harm, and a duty to guard against general harm—the warrant for the protective nature of the "fences around the Law," is the same as the warrant to place a rail in high places, a duty to protect against "evil occurrences . . . it is inappropriate for an intelligent person to endanger himself (see Deuteronomy 24:19–21), and this commandment applies everywhere and at all times, amongst men and women."[19]

There are several categories and classes of *shomrim*—the *shomer* may be paid or unpaid, or else the object may be rented or borrowed (pp. 179–80). But Freedman says that there is one form of liability that is common to all, *p'shia*, negligence (p. 180).[20] The Mishnah notes that the standard for negligence is "the custom of caretakers" (*Sefer Hachinukh*, cited by Freedman, 1999), in other words, a reasonable-person standard for that society—more care in a situation of high risk, less in a situation of lower risk (p. 182). Unlike common law, there is no right to information, but a patient has the duty to be informed to be a proper *shomer*. Where the right can be waived, a *shomer* cannot waive the duty without the loss of the entire obligation (p. 183). This is the only way to meet her proper obligation—to then "consent to that form of treatment (or non-treatment) that the patient finds is best suited to the obligation of caring for the body." If that cannot be accomplished with competence, the *shomer* cannot be a proper *shomer*—not for the body, nor for any other reason.

And so, too, the self is also the guarded entity. In this sense, Freedman is concerned primarily with the difficulty and uncertainty inherent in medical care (it is the trouble with the embodied world)—diagnoses are flawed and shifting, doctors have seven minutes to see patients in most clinical encounters, doctors disagree. There may be too much information, too little that is known, or conflicting information. There may be choices only affected by particular life situations about which the patient is the best informant—hence her duty to act as a *shomer* is critical. Unlike common law, there is no right to information, but a patient has the duty to be informed enough to be able to be a proper *shomer* over her body, hence, unlike common law, where the right can be waived, a *shomer* cannot waive her duty without the loss of her entire obligation. This is the only way to meet her proper obligation—to then consent to that form of treatment (or non-treatment) that the patient finds is best suited to the obligation of caring for the body. If that cannot be accomplished with competence, the *shomer* cannot be a proper *shomer*—not for her body, as she would not be allowed to be a *shomer* for anything else.

It must be noted that this duty is also a joyful necessity—one is a steward, not a prison guard—the body is wonderful, not degraded or sinful, life is to be exalted and protected with love. We are told that the social text for the general principle is linked to the giving of the law in Deuteronomy:

> Just take care [*hishamer*] for yourself, and take great care of your soul, lest you forget these things your eyes have seen and lest they shall depart from your heart all the days of your life, and you shall make them known to your children and your children's children. (Deut 4:9)

Hence the duty to have a duty is linked to the very notion of the family's interdependence. Freedman and others note that this takes the idea of *hishamer* —meaning to avoid harm—further, into a wider meaning, and includes one's family as well.[21]

To think of the task of being a responsible *shomer* primarily implies we must think about justice, and we must do so in four critical ways—the stewardship implies a complex world!

First: If we are serious about our Jewish bioethics, we must be serious about a steady focus on justice—nowhere is this more critical than in the evaluation of life's quality. We must think about how the wider social world shapes the intense internal world of the vulnerable patient—needing access to extraordinary care, high-intensity nursing, and serious research. It cannot be that the only request we really hear is the request to die—not ones for better staffing ratios, fully funded rehabilitation, or the technology to assist the disabled.

Second: I actually think if we can fight for only one thing in the American polity, we must fight for universal access to health care, so that robust prevention programs and early access might change how Americans think about health and illness.

Third: We need to talk about money. We have insisted and taught that families need to understand that medicine cannot always cure, which is fair enough. But "medical futility" or "poor quality of life" concerns cannot be used as a cover to mean "really expensive" and "not worth it." If we mean care is too costly for the society to bear, if this were the case, then it seems fair to ask for sacrifices all around. For example, we might ask why many physicians or hospital administrators making the claim, not to mention insurance or pharmaceutical company CEOs that make the argument for costs, make salaries that exceed that of the president of the United States.

Fourth: Someone must hear the narratives of patients. Medicaid wheelchair reimbursement was reduced in the new federal budget—fear of scams, and in California, the support services for the sickest children—CCCS—were cut deeply. Note the obscenity of this in a country that spends thirty-two billion dollars on pet care annually. Someone must hear the narratives of the disabled community and listen carefully to their fears, and real choices for good lives must be offered.

My use of the metaphor of guardianship is an extension of Freedman and draws on many deeply held arguments in both classic and contemporary Jewish thought, yet it must be stated strongly as an alternative in the large civic debate about religion, ethics, and public life, where we are confronted again and again with dilemmas in bioethics just at the junction when family choices become talismans for the way we speak of death, care, and commitment to vows. Scholars of Jewish ethics can allow the arguments of Jewish ethics to re-examine the contended ground of bioethics discourse on the problem of autonomy and its limits, the law and its aspirations, and families and their brokenness. We are guardians, not adversaries at the bedside. We are *shomrim* for the future, for one another, for the land, for the Law. We are stewards in loving duty toward the vulnerable and toward the precious. Extrapolation of this idea creates a far richer terrain for understanding a Jewish contribution toward the debate. Ontologically uneasy and epistemically uncertain, Jews live in exile, and we are in exile on these several levels—from the Garden of Eden, from the Land, and from Western modernity, in a sense.

For exiles, the entire metaphor of guardianship and of "setting watchmen at the walls" is based on a genre of waiting. In part, this is enacted ritually at many junctures in Jewish liturgical life (at Passover, waiting for liberation, on Shavuot [Pentecost], waiting for the Torah) with the goods packed, the texts in hand. We wait for a change in awareness, hope, station. The idea of the *shomer* allows us access to this fundamental moral locale of the moral subject—uncertainty and faith.

REFERENCES

Abrams, Judith (1998), *Judaism and Disability: Portrayals in Ancient Texts from the Tanach through the Bavli*. Washington, D.C.: Gallaudet University Press.

Bleich, J. David (1983), "Mental Incompetence and Its Implications in Jewish Law," in his *Contemporary Halachic Problems*, Vol. II. Ktav Publishing House, Yeshiva University.

——— (1991), "A Physician's Obligation with Regard to Disclosure of Information" in *Medicine and Jewish Law*, ed. Fred Rosner, Northvale, NJ: Jason Aronson.

——— (1998), *Bioethical Dilemmas: A Jewish Perspective*. Ktav Publishing House.

Borowitz, Eugene (1990), *Exploring Jewish Ethics: Papers on Covenantal Responsibility*. Detroit, MI: Wayne State University Press.

——— (1992), "Autonomy and Community" in *Autonomy and Judaism: The Individual and the Community in Jewish Philosophical Thought*, ed. Daniel H. Frank, Albany, NY: State University of New York Press, 9–20.

Buchanan, et al. (1990), *Deciding for Others: The Ethics of Surrogate Decision Making*. Cambridge, England: Cambridge University Press.

Committee on Bioethics (1995), "Informed Consent, Parental Permission and Assent in Pediatric Practice," *Pediatrics*, 95(2), 314–17, <http://www.cirp.org/library/ethics/AAP/> (File revised 2 January 2005).

Dorff, Elliot (1998), *Matters of Life and Death: A Jewish Approach to Modern Medical Ethics*. Philadelphia: Jewish Publication Society.

Dresser, Rebecca (1990), "Relitigating Life and Death," 51 *Ohio State Law Journal*.

Engelhardt, Tristram, H. Jr. (1999), "Whose Religion? Which Moral Theology? Reconsidering the Possibility of a Christian Bioethics in Order to Gauge the Place of Religious Studies in Bioethics" in *Notes from a Narrow Ridge*, Hagerstown, MD: University Publishing Group.

Faden, Ruth, et. al. (1986), *A History and Theory of Informed Consent*. New York: Oxford University Press.

Fisher, Celia (2002), "A Goodness-of-Fit Ethic for Informed Consent," *Fordham Urban Law Journal*. 30 (1), 159–72.

Freedman, Benjamin (1999), *Duty and Healing: Foundations of a Jewish Bioethic*. New York: Routledge Press.

Gibbs, Robert (1992), "A Jewish Context for the Social Ethics of Marx and Levinas" in *Autonomy and Judaism: The Individual and the Community in Jewish Philosophical Thought*, ed. Daniel H. Frank, Albany, NY: State University of New York Press, 161–92.

Hauerwas, Stanley (1986), *Suffering Presence: Theological Reflections on Medicine, the Mentally Handicapped, and the Church*. Notre Dame, IN: University of Notre Dame Press.

Jakobovits, Immanuel (1959), *Jewish Medical Ethics*. Bloch Publishing Co.

Kass, Leon (1985), *Toward a More Natural Science: Biology and Human Affairs*. New York: The Free Press.

Levinas, Emmanuel (1990), "The Pact," ed. Sean Hand, *The Levinas Reader*. Cambridge, MA: Blackwell Publishers.

Locke, John (1960), *Two Treatises of Government*, ed. Peter Laslett, Cambridge, England: Cambridge University Press.

Meier, Levi (1986), "Filial Responsibilities to the 'Senile' Parent: A Jewish Approach," in *Jewish Values in Bioethics*, ed. Levi Meier, New York: Human Sciences Press.

Rawls, John (1972), *A Theory of Justice*. Cambridge, MA: Harvard University Press.

Rosner, Fred (1979), "Judaism and Human Experimentation" in *Jewish Bioethics*, ed. Fred Rosner and J. David Bleich, Hoboken, NJ: Ktav Publishing House.

——— (1991), "Communicable Diseases and the Physician's Obligation to Heal" in *Medicine and Jewish Law*, ed. Fred Rosner, Northvale, NJ: Jason Aronson.

Rousseau, Jean-Jacques (1983), "Discourse on the Origin and Basis of Inequality Among Men" in *The Essential Rousseau*, trans. Lowell Bair, Plume Publishers.

Seeskin, Kenneth (1992), "Autonomy and Jewish Thought," in *Autonomy and Judaism: The Individual and the Community in Jewish Philosophical Thought*, ed. Daniel H. Frank, Albany, NY: State University of New York Press, 21–39.

Taylor, Charles (1989), *Sources of Self: The Making of the Modern Identity*. Cambridge, MA: Harvard University Press.

The Belmont Report (1979), National Commission on the Protection of Human Subjects.

Urbach, Ephraim E. (1988), *The Halakhah: Its Sources and Development*, trans. Raphael Posner, New York: Sure Sellers Inc.
Zoloth, Laurie (1999), *Health Care and the Ethics of Encounter: A Jewish Discussion of Social Justice*. Chapel Hill: University of North Carolina Press.

NOTES

1. Informed consent cases create a long historical narrative, beginning with the case of Karen Ann Quinlan. Such a narrative can be found in many places. Faden, et al. (1986) or Buchanan, et al. (1990), summarize the key cases.

2. National Commission on the Protection of Human Subjects. *The Belmont Report*, 1979.

3. See generally *Bouvia v. County of Los Angeles*, 195 Cal. App. 3d 1075, 1086 (1987).

4. See generally In re Quinlan, 355 A.2d 647, 650 (N.J. Sup. Ct. 1976).

5. See generally *Cruzan v. Dir., Mo. Dept. of Health*, 497 U.S. 261, 292 (1990); *Barber v. Superior Ct.*, 147 Cal. App. 3d 1006, 1015 (1983).

6. Fisher, Celia, Informed consent cases presented at the Fordham School of Law conference in January 2003, New York. Fisher, Celia (2002).

7. Full disclosure, I did sign such a brief, to block the action of the governor of the state in his intervention.

8. Maimonides, Hilkhot Edut 9:9.

9. Rashi, commentary on BT *Hagigah* 3b, as cited in Bleich, ibid., p. 288.

10. BT *Kiddushin* 31b. Cited and discussed at length by Levi Meier, (1986).

11. See generally Rousseau (1983), 125, 127–28; Locke (1960), 160–61.

12. See generally *Wade v. Roe*, 410 U.S. 113, 153 (1973); *Griswold v. Conn.*, 381 U.S. 479, 485–86 (1965).

13. See Exodus 18:14–20, 22, discussing the system of judges that existed after the Jews left Egypt, directly followed by the giving of the Law at Sinai.

14. See generally *Wade v. Roe*, 410 U.S. 113, 153 (1973); *Griswold v. Conn.*, 381 U.S. 479, 485–86 (1965).

15. From the *Song of Songs*, attributed to Solomon, this long narrative poem is used here for two reasons—first, because it suggests the link between love, family, and guardianship, and second, because it is read during the Passover ritual liturgy on the second day of the holiday, in the context of making covenantal promises of liberation.

16. Rabbi Saul Berman has made this point repeatedly in his (largely oral) work, first arguing this point in 1985, at a San Francisco forum on Jewish law and medicine. Also see Dorff, 1998.

17. *Kaiser Aetna v. U.S.*, 444 U.S. 164, 177 (1979).

18. A point repeatedly made in Deuteronomy.

19. *Sefer Hachinukh*, attributed to R. Aharon HaLevi, as cited by Freedman (1999), p. 139.

20. Citing BT *Bava Metsia* 42a.

21. As was noted by Noam Zohar in his close reading of this very text.

4

Do the Qualities of Story Influence the Quality of Life? Some Perspectives on the Limitations and Enhancements of Narrative Ethics

William Cutter

The place of "narrative" in ethical discourse bears examination these days, especially within Jewish communities of interpretation. Although some believe that the term may have become overused, or that the practice of seeing reality in narrative terms may cheapen our sense of logic, the role of "narrative thinking" as an illuminating contrast to formal ethical systems continues to build a serious stable of support.[1] Surely Jews have used story in plenty of ways, but the literature on literature—thinking about how narrative works—is probably not so well known to many Jewish thinkers. My essay explores several ways in which stories can be brought to bear on the Jewish bioethical conversation.

A simple enunciation of the functions of narrative in moral discourse may be taken from Martha Nussbaum in what looks like an innocent verity: "We need to involve emotional as well as intellectual activity and give a certain type of priority to the perception of particular people and situations, rather than to abstract rules." (Nussbaum, 1990, p. ix)

In the medical and healing worlds there is a flurry of narrative writing—not in the ethics domain per se—by physicians like Jerome Groopman, Oliver Sacks, Sherwin Nuland, and Rachel Naomi Remen (Nuland, Sherwin B., 1995; Sacks, Oliver, 1995; Sacks, Oliver, 1998; Remen, Rachel Naomi, 1996), along with the notable appearance of the journals *Literature and Medicine* founded by Rita Charon, and *The Bellevue Literary Review* of

Danielle Ofri. As obvious as Nussbaum's comment seems to be, the development of this "narrative turn" is note-worthy in both the therapeutic and ethical contexts. Considering the interest in narrative theory in the academy at large and within the Judaic studies environment in particular, it is time to look at narrative theory in the context of Jewish discourse on bioethics. This examination may be especially appropriate for issues of "quality of life," which, because they are intrinsically ambiguous, may be a bit resistant to abstract rules and amenable to the more flexible discourse suggested by narrative.

At this early point in my essay, I want to demonstrate one obvious link between narrative discourse and the quality-of-life discussions within this book. How seriously shall Jews take the question of male sexual performance? American television advertisements for antidotes to erectile dysfunction "frame" their sales as marked for pure enhancement, embracing stories of people romantically longing for that extended weekend or suggesting that one of the ED drugs might make for a weekend change in plans. The ads have narrated recreation over problem solving, and recreation is one story of America's concern with sexuality. But what the ads do not narrate is the story that takes place in the privacy of the doctor's office, with an anxious patient, a distraught spouse, or possibly a marriage in trouble. The fact that this selfsame "enhancement" drug might provide important therapy is missing from the narrative of the drug's value. In America we have nearly reduced Viagra or Celea and their equivalents to a certain kind of metaphor. To what extent, in reality, is the drug primarily recreational? To what extent shall we consider it therapeutic in our deployment of limited resources? Who is entitled to receive this drug should it become a rationed item? (It is already rationed—though sometimes generously—on many health plans.) And where does the quality of life sit on the continuum between recreation and therapy?

A more obviously weighty and perhaps life-saving instance might be the case of orthopedic or cardiac intervention for aged people. There may be countries where custom and insurance arrangements do not allow easy access to surgery or to such routine interventions as hip replacements for patients over a certain age. (This could include any situation where the "benefit" is limited by its utility over time or because of the mitigating reality of side effects.) But in the United States such surgery might be considered a therapeutic necessity, both by custom and by our sense of the American healing story. Clearly our national narrative of "can do" and of overcoming all adversity has a fundamental influence on the clinical disposition of cases and on their categorization. Indeed, the story that Americans tend to tell about themselves includes entitlement, success in overcoming obstacles, and the importance of pleasure and life enhancement. And where do Jews fit in this discussion? Shall we adopt the American "can do" story? Or shall we push a change in our priorities, and take up a more "prophetic" (or oppositional) voice against certain givens in the distribution of health care? It is hard to be certain how useful the term "narrative" is in this precise situation. But let us at least say that narrative suggests the playing out of the core values of American culture. While Jewish bioethics will not specifically be framed by that story or "mythos," thinkers who are asked to assist people in making decisions in the hospital need to be aware of the influence of that tradition upon patients and their families. The stories of America and Jewish American sociology have a lot in common and probably have a lot to do with the way most Jews seem to think.

So these are my framing questions. I must now set about the harder work of thinking about narrative and ethics and what the conjunction of these terms might have to say to Jewish bioethicists.

There are two ways in which one may link narrative to ethics. We may look at narratives from an ethical point of view, or we may look at ethics from a narrative point of view. Either kinds of examination—ethical situations within or ethical conclusions drawn from narratives, on the one hand, and the narrative methods of "doing" ethics, on the other—have been present even in much conventional bioethics. Any professional ethicist has in one way or another drawn on narratives or used narrative principles, especially in the first way. It is not as if these questions are entirely new for bioethics. The second "way" of involving narrative leads to a more methodological quest: How do stories work? What do they achieve? (What is now stylishly called a meta-level.) And that is the perspective that will be more prominent in this chapter. As part of the larger discussion of narrative, my chapter will discuss some epistemological questions as a component of making ethical judgments.

I must first address the more conventional role which narrative has played in ethics. When we see narratives or stories through an ethical prism, we tend to ask what particular characters ought to be doing in a story "at hand." We can then modify some of the conditions within the story, and ask whether altered conditions should or should not affect a particular character's action. Most case histories employed in bioethical discourse use narrative in this, quite legitimate, way. The approach helps readers make choices by viewing the circumstances of their choices through various lenses. The value gets played out in the story so that simulation of consequences occurs. The entire notion of stories as presenting choices, helping the individual think through consequences, and even occasionally hearing the moral voice of the author through one of the characters has been part of a long and revered tradition. It has been considered by conventional narratively inclined writers on ethics such as Wayne Booth (1987), John Gardner (1978), and Robert Coles (1989). In his discussion of the interaction of the norms that we identify with Kantian tradition and the narrative in which those norms are embedded, Robert Cover referred to alternative choices as "the counterfactual proposition" (Cover, 1995, 110). Every ethical situation is at the very least a "dilemma," meaning a situation in which one might make two readings. In our discussion about quality of life, part of the inner moral debate might be about how to define the nature of a medical decision: Is it a quality-of life-decision, or is it a critical decision for the true medical welfare of a patient or of society? Quality-of-life decisions, then, may present people with a double dilemma.

But when we view ethics from a narrative point of view, in order to gain new tools for considering our dilemmas, we are less likely to be seeking particular solutions or insights with regard to particular problems. This has not been an activity within the Jewish discussion, and remains a minority approach in general. Yet many thinkers have urged greater attention to principles of story telling in both the resolution of ethical dilemmas and in the broader epistemological sense of how to look at experience. I am thinking particularly of Stanley Hauerwas, Jerome Bruner, and Clifford Geertz, although only Hauerwas is, technically, an "ethicist." Hauerwas was—already in 1977—coaxing his readers away from what he called "standard account thinking" (in which stories played a relatively small part in ethics deliberations) toward a more frequent use of story telling methods in thinking about ethics (Hauerwas, 1997).

Jerome Bruner, our most prominent bridge between epistemology and modern educational psychology, has argued that there are two irreducible modes of cognitive functioning. Each provides a way of organizing representation in memory and of filtering the perceptual world. One seeks explications that are context-free and universal, the other seeks explications that are context sensitive and particular (Bruner, 1991).

Clifford Geertz had brought to the public a similar contrast in his foundational essay "Blurred Genres." There are two ways of explaining events, he notes: "Many social scientists have turned away from a laws and instances ideal of explanation towards a cases and interpretations one" (Geertz, 1980). So Bruner, Geertz, and Hauerwas have been with us a long time, and the tension that their emphases have set up remains with us. There is hardly a field in which something like the inductive vs. deductive process, the formal vs. heuristic design for research or some educational epistemology isn't related to this intellectual battle. It is no less present within the field of ethics, whose dominant framework has been a chiefly Kantian notion of ethics as defining general rules and applying them to particular cases. In rabbinic discourse one may begin with the narration of a case, but the conclusion to the problem presented by the case will be found in a prior decision and in the logic that one extrapolates from that decision.

But even those who, like me, prefer to organize experience in the more inductive, exegetical sense will admit that such approaches are troublesome. The approaches may add moral texture to a problem, and they may add an important element of self-reflection, but in many instances a satisfactory conclusion may have to be reached outside of the narrative. (For instance, having sympathy for a patient in a given situation is not an adequate reason for taking a particular controversial measure.) As the daughter of a friend of my family's says about her heuristically inclined father: "If you want to know how a clock works, ask Daddy; but if you want to know what time it is, ask Mom."[2]

I was drawn to Bruner and Geertz in this regard because my career was—for many years—involved with the field of educational theory and curriculum development. Parallel developments within the field of education dominated my early thinking as a curriculum specialist. I studied with thinkers who exposed the intrinsic fallacies within more common pedagogical models. Their models were more interesting than the goal-setting, objective-pursuing assessment models, but the educational culture had to fall back on goal setting and assessment in order to create conditions for hiring and firing and for evaluating student progress in some standardized way. Applied to our task here, one must say that there is a case before us, and we must make a decision; there is a policy waiting to be addressed, and we need to come up with our attitudes toward that policy. The world of decision making does not wait for narratively inclined people, whether in the field of education or in the field of ethics.

But there is a time for looking at how we think, at how our clock works, and for thinking beyond our desire to take particular actions. In Jewish bioethics substantively new methods do not appear frequently, and I hope that an examination of "narrative" approaches to problems may point the way for some new methodological directions. I am hoping, in other words, that the workings of the clock (how narratives work) may eventually influence particular decisions about particular situations, and that the cases and interpretations mode of thinking will lead beyond theory toward practice. More significantly for the project of this volume, it may actually be here that we will find some new language for conducting quality-of-life discussions, where the moral climate of Jewish thinking may be more significant than the need to act in particular cases. Arthur Frank, one of our contemporary health narrative theoreticians, refers to this as thinking "with" stories, rather than thinking "about" stories. I believe that Frank's formula is analogous to my two ways in which ethics is involved in narrative. It may be that narrative thinking is more compatible with motivational concerns as an added value to narrative approaches. If I understand Frank correctly and thoroughly enough to link one thought to another, I would suggest that narrative thinking "about" stories may fit Howard Brody's notion of thinking about how to get done what we believe needs to get done (Frank, 1995, 207).

I write here as one newly arrived to bioethics discourse. I have long studied the work of the authors in this volume and in other important volumes that have tried to bring Jewish concerns to bear on the decisions that I have helped families face in the clinical setting. I have learned that, sometimes, traditional *poskim* are more attentive than many doctors to developments in technology and science. But I write as a person from another technological universe, trained in literary issues, a reader of poems and stories, and I write as one who regularly visits the sick and their families. I have had to understand hundreds of patients in their particularity and within the uniqueness of their messy environments. Bruner and Geertz, coming from epistemology more than from ethics, have made more sense to me as a pastor who may be helping a family of six members through the maze of end-of-life decisions, or a childless couple deciding on one or another controversial method of conceiving. Once in a while a quality-of-life concern will surface ("Mama wants a face-lift"), and my liberal prophetic streak will insinuate itself and send off alarm bells: "Is this how a Jew should spend her/his money?" In most of these instances, the narrative nuances become foregrounded because the specific action may actually be less important than the family relationships, the family's history and their moral future, or their commitment to the agent's interests. And, of course, I care deeply about the implications of actions for public policy and how that policy comes to affect our national narratives. Even in the most specific ethical dilemmas, quality of life enters the equation, quality of life, at least, for certain of the parties in a situation. And in almost every case, from at least the perspective of commonsense Western ethical behavior, more than one right decision is possible.

Within the human imagination—and in particular as I have often encountered it in Jewish settings, where the tradition of *halakhah* seems to incline agents at the outset to strict judgments—the connection between narrative and ethics may suggest a modification of strict positions, or even rigorous thinking. I am speaking of the fact that once one knows the patient's "story," one may tend to soften one's point of view. What you once thought was right—in the abstract sense—needs modification in light of the concrete reality, or in light of one's own subjective experience of a situation. These are situations in which we may become convinced of the barrenness of abstraction.

I would argue that knowing the narrative of people has helped many rabbis respond more tolerantly to stickier questions in our current culture, such as sexual orientation, some instances of intermarriage, and definitions of family law and mental illness, domestic violence, and alcohol and drug abuse. Rabbis have often modified their positions once they experience the narrated reality of a dilemma which they might have judged more narrowly from a purely abstract and halakhic point of view. Narratives have prevailed to make points, either those supporting norms or those subverting them. If the *ma'aseh* I will draw on below is an example, I will be suggesting that a *ma'aseh* is often used in the Talmudic literature to trump a principle or expectations that seem to have been established. (In Bruner's terms, the story undermines the steady state. I am particularly grateful to Elliot Dorff and Rachel Adler for confirming my intuition about this.) But even when this occurs, the practice of the reader is usually to rely on the first use of narrative: the story as influencing a decision.[3]

But the second use of narrative must now be attended. Jewish theoretical work in narrative and its components has become especially interesting in recent years. David Stern's treatment of the *mashal* (parable), for example, and Jeffrey Rubenstein's work on the talmudic story examine various ways in which stories have been utilized within rabbinic literature. Stern's discussion provides the basis for our transition into the second way of examining the relationship between narrative and ethics. Stern points to three ways in which parables have been used, with only one of these ways being restricted to the function of the exemplum (that is, story used to teach a specific ethical point). What is central for our concern is Stern's contrast between the parable's communicating function and its singular literary features. He emphasizes that we should not eliminate these literary features of the *mashal*, its more expanded poetic function, if you will, on account of our urgency to support the communication function of the material.[4] This "communicative function" is what pushes narrative to participate in helping us make specific decisions, but it diverts us from understanding the "story-ness" that is before us. (In terms of my earlier quip, "how the clock works.") Rubenstein's contributions include his insistence that we read talmudic stories within the larger context of the Gemara, including the *sugya* (the complex thread of talmudic discourse) at hand, and the parallel sources and traditions that were known to the talmudic editors and their audience (Rubenstein, 1999). Combining the promptings of Stern and Rubenstein facilitates the transition from narrative as the source of ethical deliberation to narrative technique as a way of helping us discuss ethics. In other words, I would argue that some of the best work on rabbinic literature (and I could add others like David Kraemer, Aryeh Cohen, and Daniel Boyarin along with a host of textual inquiry scholars)[5] has a significant relationship to my earlier discussed appeal from Hauerwas, Bruner, Geertz, and Nussbaum. In this work one expands one's sight beyond story as a way of solving one dilemma to a way of looking at the world.

I will move now to take up a famous rabbinic story that came to be used and has often served as an "exemplum," or at least a "locus classicus," for an important principle with regard to end-of-life issues. I will use the story, even more vigorously, to suggest an examination of "story-ness." It is the tale of Rabbi Judah's servant (BT *Ketubbot* 104a).

> On the day when Rabbi (Judah) died, the rabbis decreed a public fast and offered prayers for heavenly mercy. They, furthermore, announced that whoever said that Rabbi was dead would be stabbed with a sword.

> Rabbi's servant ascended the roof and prayed: "The immortals desire Rabbi to join them, and the mortals desire Rabbi to remain with them; may it be the will of God that the mortals may overpower the immortals." When, however, she saw how often he resorted to the privy, painfully taking off his tefillin and putting them on again, she prayed: "May it be the will of God that the immortals may overpower the mortals." As the rabbis continued their prayers for heavenly mercy, she took up a jar and threw it down from the roof to the ground. At that moment they ceased praying and the soul of Rabbi departed to his eternal rest. (Soncino Translation)

Here the prayers of Rabbi Judah's disciples contend with the prayers of those on "high" and are apparently prevailing to keep him alive. The prayers can be defined, in the later halakhic language, as an impediment (a "*mone'a*") to the dying process. Rabbi's servant, at least in the story, is the only person who truly experiences her master's pain, a fact which is developed through narrative devices and linguistic figures which point to her master's pain and its frequency. When she is finally convinced by understanding his suffering to do something about the nonstop prayer intervention (the *mone'a*), she breaks a jar, the worshippers on earth are distracted from their prayers, and Rabbi's soul is allowed to depart in that split second. Formally, then, there has been a "turning away of the impediment to death" (a "*mesir mone'a*"), something has taken away the impediment to death's natural course. In technologically developed cultures the principle has been extended by many *poskim* to mean that artificial technological means for extending life may be removed. The story of Rabbi Judah's servant suits Robert Cover's suggestion that "every narrative is insistent in its demand for its prescriptive point" (Cover, 1995, 96). But it is also a "narrated event" and therefore contains further propositions through a kind of "germinative power."

Indeed, we sense that this tale generates separate events or separate awareness from which we may extrapolate. What are they? The story relates the affection of a servant for a master, which has some influence on the decision made in this dire circumstance. This could also be seen as a story about how wrong community norms or expectations may be, since, if read as an exemplum, the story demonstrates a subversion of the conventional wisdom at the beginning of the story: we all want Rabbi to live. If some stories in the Gemara are often used to "trump" the philosophical logic or the traditionalist expectations of the discussants in a case, the story of Rabbi Judah's death would be a case in point, (along with some other stories: see note 3). Among other narrative elements that the story reveals, the story pays attention to the condition of time in influencing the servant's response.

Narrative and its principles are the bridge between abstraction and actual experience, and narrative is always anchored in character or characters. The servant is the only person in the narrative who pays attention to the suffering person himself—that is, not as an abstraction. Modern health narrators like Groopman and Nuland have made the person, rather than the illness, the cornerstone of their discourse, but it is especially gratifying to see this concern articulated by Yeshayahu Leibowitz in "The General and the Particular in the Theory of Medical Practice" (Leibowitz, 1982). Yet ethical deliberation, like the halakhic tradition drawing on this story, has generally privileged the situation—the illness—and the abstract principles behind the situation. Reading narrative as a way of looking at ethics may prompt a more patient-centered reading of a dilemma.

I would like to add some aspects of narrative theory to the Jewish notion of covenant. David Ellenson (taking a cue, in his own words, from Irving Greenberg and David Hartman) posits autonomy as a core value and an essential part of being a partner in Covenant, something which Dan Brock argues in his contention that people have the right to make choices that may run counter to normative bioethical thinking. In other words, the very notion of Covenant, in liberal terms, assumes certain autonomy on the part of an agent (Ellenson, 1991, Brock, 1993, 110). There may be a Bakhtinian sense here of inter-subjective alliance, and even the possibility of having to behave against one's conscience in certain situations (Zohar, 2003). Social or dialogical alliance may also be more likely to open up "quality of life" discussions. Narrative thinking, per se, may make the case that there are many possible conclusions and not simply the dramatization of options from which one must choose the best alternative.

Can it be that systematic ethics and halakhic formalism seek to free agents from the pain of the tragic, incomplete nature of decision making? In other words, might it be that narrative thinking removes the protection supplied by classic halakhic formalism along with the consistency of philosophical logic? I will argue below that the compromises of decision making may have an almost "allegorical" relationship to the way in which we have to deal with limited economic and social resources (Zoloth, 1999). I will now suggest that there exists a relationship between the flawed nature of many ethical choices and the limits inherent in literature and language.

J. Hillis Miller has argued that the relation between ethics and story has to do with our limited epistemological capabilities and the need to renounce one interpretation of a narrative in favor of another (Miller, 1987). I will suggest that Miller's thinking has cognates in at least one Jewish tradition—the one tradition that is most interested in narrative theory.

For this phase of my discussion I begin with a principle from Rabbi Nahman of Bratzlav, articulated most fully in a number of specific homilies (especially, Alef of Likkutei Moharan, para. resh/lamed/dalet). His sense of story, it has seemed to me, blends intuition with a startling theoretical sophistication. The term *tsimtsum*, perhaps too well known and too loosely known among contemporary readers, signifies that contraction of God which makes possible the creation of the world by permitting (perhaps creating) finitude. Yet, by removing the infinite light, or the light of the infinite (known as "*or ein sof*") from within, the tragic cycle of the full human experience is installed within human experience. For Nahman, story is the embodiment of that ebb and flow of divine light, a fullness and imperfection that is the human condition. One needs to know how to tell a *ma'aseh* (a story, and perhaps a special kind of Bratzlav story) because in every story there is a contraction that mirrors the contraction necessary for creation. Whenever we want to purify a thought by way of a story, we create a process that includes purifying the story through thought. It is a recurrent cycle of abstraction of the point from a story and the re-situation of the point within the narrative. What Nahman seems to be suggesting, to place this in somewhat reductionist terms, is an event of compromise—an ontology of incompleteness that in turn forces an incomplete epistemology: our finite competence to understand the precise nature of the ontological condition. Kathryn Hunter more recently noted, citing George Lukacz: "Like theory, plot is the mediating force in the dialectical movement from concrete reality to abstract representation and back to conscious participation in reality" (Hunter, 1991, 65). Nahman's homilies suggest that stories create their own meanings and their own multiple interpretations out of their very nature.

I don't want to make a case here for a direct connection between Cover's theory and kabbalistic—much less Hasidic—metaphysics, but Nahman's attachment to story does provide the concretization of the abstract principle before moving the reader to another abstract principle that grows out of the concrete action of narrative. A similar interplay must always be with us when we consider bioethical dilemmas. Certain kinds of anthropologists, it seems to me—(following Bruner and Geertz) may not build stories out of data so much as they discover data because of the stories that shape their perceptions (Bruner, cited in Mattingly and Garro, 2000, 1–49). Nahman's sense of our limited epistemological competence is precisely what drives the "ethics of reading" that some post-modernists like J. Hillis Miller have promoted: every reader makes an exegetical judgment that is an allegory of the kinds of choices that characters make within a story; and for the agents of our bioethical dilemmas, the use of narrative to frame new and complex questions that actually add to the pool of interests about which we must compromise creates an important alternate way of looking at medical situations which are inherently compromised.

Narrative, in sum, contextualizes all discussions of norms; it surprises with its conclusions, it places the characters at the center of our attention. Narrative respects a variety of perspectives, including that which is influenced by the broader culture in whose context an ethical decision is made. It insists upon the overthrow of the steady state or its subversion and it introduces time as a factor in decision making. And narrative somehow stimulates a counter-narrative. As Arthur Kleinman said recently of Susan Sontag, "A good depiction may force us to argue back" (Kleinman, 2003).

According to Adam Zachary Newton, whose work has been essential to my thinking here, the ethics of reading narrative requires us to think the infinite and the transcendent, but to live in the finite (Newton, 1995). It gives us grounds along with the knowledge that we must all make tragic and incomplete resolutions of our dilemmas.

The imperfect world in which we live must envision an ideal and work with the real, where an entire civilization is now able to ask quality-of-life questions that would have been meaningless before the seventeenth century. Those questions will not always lead to ideal solutions. While I can't imagine that traditional halakhic formalists would argue against the existence of this tension between the ideal and the real, I do suggest that narrative thinking opens new possibilities for this discourse that have been ignored heretofore.

Surely the Jewish tradition offers an opening to narrative thinking about quality-of-life questions, even as it has had an appropriately troubled relationship to our twenty-first-century quest for quality of life. Narratives are most often about troubled relationships.

REFERENCES

Booth, Wayne (1987), *The Company We Keep, An Ethics of Fiction*, Berkeley: University of California Press.

Brock, Dan (1993), "Quality of Life Measures in Health Care and Medical Ethics," in *The Quality of Life*, ed. Martha Nussbaum and Amartya Sen, Oxford: Clarendon Press, 95–132.

Brody, Howard (2000), *Stories of Sickness*, Oxford: Oxford University Press.

Bruner, Jerome (1991), "The Narrative Construction of Reality." *Critical Inquiry* 18/1.

Coles, Robert (1989), *The Call of Stories*, Boston: Houghton Mifflin.

Cover, Robert (1995), "Nomos and Narrative," in *Narrative Violence and the Law: The Essays of Robert Cover*, ed. Martha Minow, Michael Ryan, and Austin Sarat, Ann Arbor: The University of Michigan Press, 95–172.

Ellenson, David (1991), "How to Draw Guidance from a Heritage" in *A Time to Be Born and a Time to Die: The Ethics of Choice*, ed. Barry S. Kogan, Hawthorne: Aldine de Gruyter, 219–232.

Frank, Arthur (1995), *The Wounded Storyteller: Body, Illness and Ethics*, Chicago: University of Chicago Press.

Gardner, John (1978), *On Moral Fiction*, New York: Basic Books.

Geertz, Clifford (1980), "Blurred Genres: The Refiguration of Social Thought," *American Scholar*, No. 49, Spring.

Hauerwas, Stanley (1997), *Truthfulness and Tragedy: Further Investigations into Christian Ethics*, with Richard Bondi and David E. Burrel, South Bend: University of Notre Dame Press.

Hunter, Kathryn Montgomery (1991), *Doctors' Stories: The Narrative Structure of Medical Knowledge*, Princeton: Princeton University Press.

Kleinman, Arthur (2003), A review of Susan Sontag, "Regarding the Pain of Others," in *Literature and Medicine*, 22/2, New York: Farrar, Straus and Giroux.

Leibowitz, Yeshayahu (1982), "The General and the Particular in the Theory of Medical Practice," (Hebrew) in *Emunah, Historiah VeArakhim*, Tel Aviv.

Mattingly, Cheryl and Garro, Linda (2000), "Narratives as Construct and Construction" in *Narrative and the Cultural Construct of Illness and Healing*, Berkeley: University of California Press.

Miller, J. Hillis (1987), *The Ethics of Reading*, New York: Columbia University Press.

Newton, Adam Zachary (1995), *Narrative Ethics*, London: Harvard University Press.

Nuland, Sherwin B. (1995), *How We Die: Reflections on Life's Final Chapter*, New York: Vintage Books.

Nussbaum, Martha (1990), *Love's Knowledge: Essays on Philosophy and Literature*, New York: Oxford University Press.

Ochs, Peter and Levene, Nancy (eds.) (2002), *Textual Reasoning*, Grand Rapids: William B. Eerdmans.

Remen, Rachel Naomi (1996), *Kitchen Table Wisdom: Stories That Heal*, New York: Riverhead Books.

Rubenstein, Jeffrey (1999), *Talmudic Stories: Narrative Art, Composition and Culture*, Baltimore: Johns Hopkins University Press.

Sacks, Oliver (1995), *An Anthropologist on Mars: Seven Paradoxical Tales*, New York: Knopf.

——— (1998), *The Man Who Mistook His Wife for a Hat and Other Clinical Tales*, New York: Simon & Schuster.

Stern, David (1981), "Rhetoric and Midrash," *Prooftexts* 1, 3.

——— (1991), *Parables in Midrash: Narrative and Exegesis in Rabbinic Literature*, Cambridge, MA: Harvard University Press.

Strawson, Galen (2004), "A Fallacy of Our Age—Not Every Life Is a Narrative," *Times Literary Supplement*, 15, 13ff.

Zohar, Noam (2003), "Cooperation Despite Disagreement: From Politics to Healthcare," *Bioethics* 17 (2), 121–41.

Zoloth, Laurie (1999), *Health Care and the Ethics of Encounter*, Chapel Hill: University of North Carolina Press.

NOTES

1. For two examples of this emphasis see: Nussbaum (1990), Newton (1995). The collection *Textual Reasoning* (Ochs and Levene, 2002) reflects this in its most contemporary terms.

2. I feel obligated, as well, to add even more substantive notes of caution, reflected in some interesting resistance to the narrative trope. See, for example, Strawson (2004). I am grateful to my colleague, Tamara Eskenazi, for this article and for many discussions of narrative thinking, for which we share an affinity.

3. For rabbinic stories that illustrate this strategy, see BT *Ketubbot* 104a, *Yevamot* 65b, and Mishnah *Bava Metsia*, 7:1.

4. Stern (1981) and more generally, Stern (1991; see especially p. 15).

5. For a sampling of leaders in this work, see: Ochs and Levene (2002), especially the articles by Laurie Zoloth, Robert Gibbs, Peter Ochs, and Tikva Frymer Kensky.

5

The Place of Hope in Patient Care: A Review Essay of Jerome Groopman's *The Anatomy of Hope: How People Prevail in the Face of Illness* (New York: Random House, 2004), xvii + 235 pages

Elliot N. Dorff

We all want quality of life. Further, we want it throughout our lives. We may differ as to exactly what that means—some think that a busy life is desirable, for example, while others want a relaxed life. Still, the vast majority of us want a life of reasonable length crowned with meaningful relationships and achievements and, except in the course of striving for some goal, relatively free of pain.

These desires do not stop with advancing age or with illness. In fact, it is especially in those contexts, when our quality of life is threatened, that we think about how we can preserve it and perhaps even enhance it. In such circumstances, we cling to a hope that we can live as long as we would like with meaningful relationships and activities to occupy our time and with as little pain as possible, both physical and psychological. The more aged or sick we become, the more fervent our hope that we will be blessed in that way and the deeper our fear that we will not.

In his engaging and wise book, Dr. Jerome Groopman describes the significance of hope in patient care and yet its dangers. Specifically, his story about Claire Allen, whose doctor told her the unadulterated truth about her illness in a frank, matter-of-fact way, indicates that doing so can rob the remainder of a patient's life of all joy and meaning. On the other hand, Henry Gold's doctor gave him false, unfounded hope, and that deprived him of the ability to speak to his family and friends about his illness and, more importantly, about his life and his hopes for them. What is clearly needed is a middle ground, in which patients learn the truth of their illness while yet learning about what could give the remainder of their life meaning.

This is important for doctors as well as patients. Groopman's account of his mentor, Dr. Richard Keyes, illustrates a phenomenon that has been well documented—namely, that when doctors become convinced that they cannot cure a patient, they stay away. Part of that is simply a rational use of time, for, especially in an age of health maintenance organizations and other arrangements that put a premium on the time spent with patients and on rates of cure, doctors who have no hope to heal a patient do not "waste" their time with him or her. On the other hand, though, this reveals a disturbing piece of contemporary medical practice, for it encourages doctors to understand their profession as if they were mechanics, with their patients reduced to bodies, such that there is reason for doctors to spend time with only those patients whom they have a reasonable chance to cure.

I first became aware of how this approach to medicine demeans both the patient and the doctor when I served on the Los Angeles Jewish Hospice Commission in the early 1980s. It was founded by David Schulman, a Jewish lawyer who serves as Los Angeles City Attorney to prosecute AIDS discrimination cases, and chaired originally by Rabbi Maurice Lamm, and one of its first members was Dr. Lawrence Heifetz, an oncologist at Cedars-Sinai Medical Center. Larry is an upbeat, almost jovial man, with a ready smile and a gleam in his eye. He is also a brain surgeon, and by his own admission at that time, some 75 percent of the brain tumors he saw he could not cure. I once asked him how he maintains such an optimistic, joyful attitude toward his profession and toward life in general while seeing death and the threat of death day in and day out. He told me that a long time earlier he decided that he would surely do his best to cure his patients, but even if he could not cure a patient, he could still care. That is what brought him into the Los Angeles group that founded the first Jewish Hospice Commission anywhere, for in hospice care the goal shifts from cure, which has been deemed impossible, to care by all involved. Doctors provide and monitor pain medication as well as interact personally with the patient; and nurses, family members, clergy, and social workers all pitch in to help the patient feel comfortable and do meaningful things. The patient is treated as a person, not a machine, and the doctor then becomes a person as well, as he or she uses medical expertise and personal warmth to help the patient physically and emotionally.

The downside of hospice care, of course, is that those who engage in it give up all hope of recovery. When I became Chair of the Los Angeles Jewish Hospice Commission after Rabbi Lamm left Los Angeles, Rabbi Immanuel Jakobovits, the author of the first extensive book on Jewish medical ethics and then Chief Rabbi of the British Commonwealth, wrote me a letter asking about that. He knew and had great respect for Rabbi Lamm, he said, but how could we allow hospice care if that entails abandonment of all hope for recovery? I responded to him that Judaism cannot be reasonably read to require unrealistic hope, for, after all, the Mishnah forbids us from praying for changing something that has already been determined (Mishnah *Berakhot* 9:3), and the Talmud declares that whenever an injury (or worse) is likely, we may not depend on miracles to remove it (BT *Kiddushin* 39b; see also BT *Pesahim* 64b). Hospice care, though, does not remove reasonable hope—for example, for minimal pain and maximal meaning and social interactions in the remainder of one's life. In that way, hope at the end of life is very much like hope at life's other stages: throughout life our ability to hope for something gives us psychological and religious meaning and thus is crucial for our well-being; but the objects of our hope must change from one period of life to another to reflect what is reasonable and appropriate for that period. (It is reasonable to hope, for example, that one's child will begin to utter discrete words and then sentences during his or her second year of life, but not earlier.) It was after that interchange that Rabbi Jakobovits permitted Jewish institutions in Great Britain to offer hospice care.

Part of the problem, of course, is that until recently doctors were trained to pay attention to only physical factors in the patient's illness. That is, American medicine did indeed treat the patient as a machine that needed to be fixed, thus unwittingly making doctors glorified mechanics. In the early 1990s, however, I was involved in a curriculum committee at UCLA Medical School that created its program called "Doctoring," a course that every medical student there must take all four years of medical school. Through a series of cases enacted by professional actors in the first two years of medical school (this is Los Angeles, after all!) and by real patients in the hospital in the last two years, students are taught to recognize and deal with the diverse non-physical aspects of patient care, including the anthropology, sociology, economics, and ethics of medicine as well as its religious and spiritual aspects.

So, for example, the first case in the curriculum concerns a fifteen-year-old female who comes to see you, the doctor, with a complaint of stomach problems. When you examine her, however, you discover that she has no stomach problems whatsoever; she is just pregnant. From a physical point of view, it would be better if she were eighteen and much worse if she were twelve, but as long as she is in otherwise good health, at fifteen she is probably going to have a normal pregnancy. But she is unmarried, Hispanic, and Catholic. All the problems with her medical care, then—problems that will significantly impact her own physical welfare and that of her fetus—are not themselves physical. What does it mean to be pregnant and unwed in the Hispanic community? Will her parents throw her out of the house? If so, where will she live, and how will both she and the fetus survive? If not, will she keep the child or abort it? As a Catholic, she is unlikely to abort the fetus, but some Catholics disobey official Catholic teaching on this. If she carries the fetus to term, will she give it up for adoption or keep it? Will she have to care for it, or will her parents do that so that she can finish high school and maybe even college? Where is the fetus's father in all of this? Doctors may not be the ones to help her deal with all these issues, but they surely must be cognizant of them and know on whom to call for help with them, for the girl's very life and that of her fetus depend on resolving these issues effectively. This new form of medical education promises both doctors and patients a more humane interaction, one that is also medically more successful. It also bodes well for assuring us that doctors of the future will know how to communicate a poor prognosis much better than the doctor did in Groopman's case of Frances Walker.

Groopman says, however, that there can be hope only if the patient has real options and choices so that the patient has some control over his or her medical situation (Groopman, 2004, 26). In this I beg to differ. On one plane, I surely do not control what happens to me after I die, but Judaism, like many other religions, holds out hope that in some way we continue to live after death. Here again, I do not control that, but I can surely hope that that is so. Jews historically have differed in what they have hoped for after our physical life. Biblical books maintain that we live on through our children and through the effect we had on others during our lives. The rabbis of the Mishnah and Talmud assert that we actually live in another form of existence after our physical death, and medieval Jewish philosophers like Saadia and Maimonides argue as to whether we have a body in that future world or not. (A good exposition of Jewish beliefs about life after death is Neil Gillman's book, *The Death of Death: Resurrection and Immortality in Jewish Thought.*) Groopman may be alluding to such convictions at the very end of his book (Groopman, 2004, 210), where he says, "There is deep comfort in the sense that we are not alone when we try to move out of the shadow of death," but I am not sure whether he has doctrines about life after death in mind or not. In any case, people do find meaning, comfort, and hope in such beliefs, even though they do not control the nature of life after life.

Groopman also notes there that patients find meaning in prayer. He especially applauds the prayer of one of his patients, namely, "I pray that he [God] helps my doctors, that he gives them wisdom" (Groopman, 2004). He also asserts, "I have found strength and solace in both the insights of tradition and the structure of ritual." Interestingly, I know of one doctor, Michael Orlow, who took time off between college and medical school to earn a Master's in Jewish Thought at the Jewish Theological Seminary of America, and he wrote his Master's thesis on the need of medical personnel to have rituals in their professional lives. He wrote, for example, a prayer that surgeons might use before doing surgery and another prayer for medical personnel to use when their patient dies. Doctors need both "the insights of tradition and the structure of ritual" as much as patients do.

Furthermore, hope involves not only medical matters, but also the plethora of factors that shape every person's life, many of which are out of a person's control. Patients often hope, for example, for human interaction, for illness is *isolating*. As Buber taught us, however, one can have an I-Thou relationship with someone else only if both parties agree to enter it. Thus if I am a patient, I do not control whether you will even come to see me, let alone have a real human interaction with me, but I can surely hope that you will and maybe even take steps to convince you to do so.

But few of us want to visit the sick, especially when they are in a hospital. First, because we are not seeing the person in the usual times and places, we have to put a visit in our calendar. We have to go there, pay for parking, find the patient's room, and then, when we get there, it is quite possible that the patient is either out for tests or sleeping. That only reinforces one's reluctance to go to the trouble to visit someone in the hospital. If the patient is there and awake, the visitor faces another set of problems, especially this one: What do I talk about? Patients and visitors tire of discussion of the food and the weather in about ten seconds.

The Jewish tradition was aware of all of these factors. That is why, first, it makes visiting the sick (*biqqur holim*) a commandment, something that God demands that we do, whether we want to or not. Furthermore, the tradition gives us some good advice about how we should act when we do visit the sick.

First (and doctors take note), we should sit down, for if we stand while the patient is lying in bed, our body language communicates that we have power and the patient does not. That is exactly the wrong thing to convey to a patient, for illness is by its very nature *debilitating*, and visitors should not reinforce that feeling in the patient.

Standing also communicates that the visitor does not intend to stay very long, and patients read that cue immediately. Since they do not want to impose on visitors any more than they already have, they will often say to standing visitors, "Thank you for coming," and assume that the visitor wants to leave quickly without engaging in any meaningful conversation.

Second, one should talk about the very things that one would normally speak about if the patient were not ill—sports, books, movies, politics, shul business, etc. Illness, after all, is not only debilitating and isolating; it is also *infantilizing*. In talking about the usual subjects that engage the patient and visitor, the visitor immediately communicates that the patient is still an adult and that his or her opinion still counts.

One of my most memorable lessons in the rabbinate occurred when I was working on my doctoral dissertation the year after I had been ordained. I was asked to do three sessions on Jewish theology for the residents of the

Jewish Home for the Aging in Manhattan. When I asked the social workers who planned these sessions why the residents wanted to study Jewish theology, of all things, they responded, "Because they are sick of Bingo!" All of the residents had a college education, and even though their bodies were no longer functioning well, their minds were fine, and they needed to be challenged. I mention this not to suggest that visitors should normally talk about Jewish theology, but simply to indicate that patients continue to need to learn and to have their minds stretched in order to feel like the adults they are.

Third, illness, especially chronic illness, is *boring* for both patients and visitors. Patients need a reason to get up in the morning. Knowing that visitors will arrive is one such reason, but an ongoing project is even more compelling. The Jewish tradition suggests one such project—namely, that visitors help the patient create an ethical will. Originally a letter that the patient wrote to his or her children, it now can be an audiotape or videotape in which the patient tells the family history, describes the aspects of Judaism that are important to him or her, expresses hopes for his or her children and other family members, and articulates love for family and friends. Creating such a document for one's family (especially one's grandchildren) gives life meaning, even when illness robs a person of the ability to do much else.

Finally, according to Jewish law one has not fulfilled the commandment to visit the sick unless the visitor prays with and for the patient (*SA*, *Yoreh De'ah* 335:4, gloss.). The prayer may be spontaneous and in English or one of the traditional, Hebrew prayers, and it need not be longer than a sentence or two. Praying with a patient invokes the visitor's and the patient's hope that God will aid the patient in coping with the disease and in making the remainder of the patient's life meaningful. A traditional, Hebrew prayer also invokes the support of the whole Jewish community, past, present, and future. Jews who are not used to praying often find prayer at the bedside to be surprisingly powerful for both the patient and visitor.[1]

Finally, one should note that faith and doubt interact in the structure of hope not only in people facing illness, but in all of us. As the French Jewish philosopher Paul Ricoeur noted, when we first learn about religion (and, indeed, about life), we believe everything we are told; we are naive. Then we begin to doubt things, until we doubt everything. But then, if we pursue matters sufficiently, we enter into a third stage, which he dubs our "second naiveté," when we know all the problems in a given belief or ritual and nevertheless affirm it, albeit in a new, more sophisticated way.[2]

When patients face illness, some are naive about it, holding hopes that unfortunately will disappoint them. Nevertheless, for some, this naive faith works to help them cope with their illness. For others, whom illness catches in the stage of doubt, illness adds to their depression and hopelessness. But for a third group, who have progressed to a faith that is aware of the likelihood of imminent death and yet affirms the value of one's remaining life, of people, and of God—those who have achieved a second naiveté—their faith can be the source of strength, meaning, and hope.

Thus while hope plays an important role in what we mean by quality of life throughout our lives, it becomes absolutely critical at the end of life. Then we must identify what we can realistically hope for and the means we can employ to achieve it. We also must locate our hopes in the deep ground of our fundamental beliefs about life and death, about the purpose and meanings of life, and about what we hope to leave behind when we die.

In the end, then, as Groopman indicates ever so tentatively at the end of his book, what on the surface appear to be straightforward, medical questions turn into broad and deep religious matters. The "lig" in the word "religion" comes from the same Latin root as the word "ligament": it means ties, connections, bonds. Among other functions, religions give us a broad, "Grand Canyon" picture of our links to the other members of our family, our community, the larger human community, the environment, and the transcendent element of human experience, imaged in the three Western religions as God. The various religions of the world understand that in different ways, and they also differ in how they depict the kind of individuals and societies we should strive to be, the nature of the good life. We human beings feel the need to explore such matters throughout our lives, but never more than at the transitions of life—birth, adolescence, marriage, and death. So physicians, nurses, other health care personnel, and all of us as patients and visitors who strive to preserve and enhance our quality of life and that of others ultimately must address not only the physical and emotional means for doing so, but the religious means as well. While Groopman himself shies away from drawing this conclusion, the real message of this book is that quality of life entails attention not only to the concrete and immediate—pain medication, food, toileting, etc.—but also to the abstract and the ultimate—the meaning of life and death, the relationships we have, the goals we strive to achieve. In pushing us to go beyond the issues of true and false hopes and how to convey them to dying patients, Groopman thus prods us to penetrate how we achieve quality of life in every period and aspect of our lives.

REFERENCES

Dorff, Elliot N. (1998), *Matters of Life and Death: A Jewish Approach to Modern Medical Ethics*, Philadelphia: Jewish Publication Society.

——— (2003), *Love Your Neighbor and Yourself: A Jewish Approach to Modern Personal Ethics*, Philadelphia: Jewish Publication Society.

Groopman, Jerome (2004), *The Anatomy of Hope: How People Prevail in the Face of Illness*, New York: Random House.

Ricoeur, Paul (1967), *The Symbolism of Evil*, New York: Harper and Row.
Wallace, Mark I. (1990), *Second Naiveté: Barth, Ricoeur, and the New Yale Theology*, Macon, GA: Mercer.

NOTES

1. For more on visiting the sick and how that affects hope, see Dorff (1998), 255–64, and Dorff (2003), chapter 7.
2. Ricoeur (1967), 351–57. For an interesting application of Ricoeur's insight to modern Protestant thought, see Wallace (1990).

II

INSIDE THE FAMILY: THE CARETAKERS' QUALITY OF LIFE

6

Balancing Parents' and Children's Quality of Life: Dilemmas in Caregiving

Dayle A. Friedman

The challenge of family caregiving for elders is ubiquitous and wrenching for many in the Jewish community. The "age wave" has hit the Jewish community more dramatically than the general population.[1] The 2000–2001 National Jewish Population Study revealed that 19 percent of American Jews are over sixty-five, compared to 12 percent of the U.S. population.[2] The fastest growing group is those over seventy-five, already 9 percent of the American Jewish community. Among Jews over sixty-five, over one-third report that their health is poor or fair; 26 percent of elderly Jewish households have a member with a health condition that limits daily activities.

This age wave thus creates a parallel *caregiving* wave. Jewish families are caring for elderly members living longer, and with more extended periods of greater dependency, than ever before. Three factors make the demands of the caregiving wave more pronounced in the Jewish community: the higher proportion of elders in our midst; the community's lower birthrate, and consequently ever smaller pool of caregivers; and the geographic mobility of the Jewish population.[3]

In the face of increasingly complex and protracted caregiving needs of elders, their children must discern what they are obligated to do, and how to balance their obligations to parents with compelling competing responsibilities, including work, children, and partners. Whose quality of life takes precedence in this harrowing juggling act, the parent's or the adult child's? Seeking guidance, we look to precedent from our tradition. Our texts are filled with accounts of exemplary, self-sacrificing deeds of filial piety. For example, in a single passage in the Babylonian Talmud, Kiddushin 31a–b, we find the following:

- Dama ben Netinah, a gentile, gave up an extremely lucrative business transaction rather than disturb his sleeping father to get to the merchandise (the key was under his father's pillow).
- The same Dama ben Netinah allowed his aged mother to strike him on the head, rip off his golden cloak, and spit in his face. Having suffered all of this, he nonetheless did not offer her reproof.
- Rabbi Tarfon allowed his mother to use him as a footstool, climbing on his back as he bent over, so that she could get into bed comfortably.

These accounts emphasize the parent's quality of life over the adult child's. Reading them, we might deduce that the adult child's quality of life counts not at all. These texts portray children sacrificing financially, physically, and emotionally, attending assiduously to their parents' quality of life, and ignoring or surrendering their own. As we shall see in our investigation, both the realities faced by today's caregivers and the values of our tradition are far more complex.

I approach the issue of our obligations to aging parents as a rabbi, a pastoral caregiver, an adult child of aging parents, and as a woman. As we explore the challenge of balancing quality of life in filial caregiving, it is useful to make some phenomenological observations.

First, caregiving for aged parents as described in Jewish tradition is equally obligatory for men and women; interestingly, the examples we find in the text above are of sons caring for parents. We can only guess whether these examples are cited because they represent the rare exception to the norm. Perhaps these cases of male caregiving attracted the male rabbis' attention, while they were hardly conscious of women's routine caregiving for elders, so expected and so of a piece with their caring for children and spouses.

We do know that the reality in our contemporary North American culture is stark: caregiving at both ends of the life cycle is nearly universally a woman's role. Men typically become primary caregivers only when women are unavailable. The burden of parent care falls disproportionately upon daughters. This inequity sharpens the dilemmas we are examining. Further, the plight of caregivers is made even harsher by the nearly universal tendency for one child in the family to become the primary caregiver, even when others are present and potentially available to share. At this end of the life cycle, as at the other end, women render care in exquisite isolation.[4]

There are therefore important gender dimensions to these questions. Some feminist ethicists have suggested that an ethic of care replace or supplement an ethic of justice. In the ethic of care, moral reasoning emerges out of the context of the particular relationship at hand.[5] Practical experience, not abstract principles or external authority, provides the basis for correct choices.[6] Using an ethic of care, we would examine the dilemmas of caregiving in terms of the "activity of care" (Tronto, 1994, p. 648), not just abstract notions of obligations. We would also consider the situation in light of the interdependence and reciprocity of the parent-child pair.[7] A Jewish ethic of care would address the needs and well-being of caregivers and care receivers alike, both in individual families and in society at large. It might well lead us to attend as a community to the expectations and demands placed on "women in the middle" of work, childrearing, and caring for parents, and might even prompt the allocation of communal resources to the support of these women.[8] A Jewish feminist ethic of care could be a fruitful resource for the caregiving wave.[9] While not yet fully articulated, we will use this approach's focus on relationships and mutuality below as we grapple with a specific case example of family caregiving dilemmas.

Now, to the issue at hand: How can our tradition guide us in navigating among the treacherous shoals of caregiving? In our exploration, we seek values that emerge from *halakhah*, from *aggadah* (Oral Torah in the broadest sense, including midrash, folk culture, and literature), and from the lived experience of the Jewish people. This broad approach to the sources is preferable to a narrow focus on halakhic discourse, which runs the risk of becoming halakhic formalism. By "halakhic formalism" I mean a method that identifies precedents from rabbinic texts in order to deduce or extrapolate norms, which in turn are thought to yield authentic Jewish prescriptions regarding specific issues.[10] Instead, we are searching out values which an individual or community would need to weigh in evaluating choices in a given situation.[11] Aggadic sources will yield values rather than norms; yet any particular source will only furnish part of the requisite range of perspectives. In fact, our tradition is much more nuanced and rich than might be suggested by the *Aggadot* cited earlier, taken alone.[12]

FUNDAMENTAL ASPECTS OF FILIAL PIETY

Clearly, filial piety is a weighty responsibility. There are two fundamental dimensions to our obligations toward parents, as outlined in Torah and rabbinic explication. First, we are called to give honor, *kavod*, to our parents.

> Honor your father and your mother, that your days may be long on the land that the Eternal, your God, is giving you. (Exodus 20:12)

Secondly, we are obligated to treat our parents with reverence, *mora*:

> You shall each revere your mother and your father, and keep My sabbaths: I, the Eternal, am your God. (Leviticus 19:3)[13]

The rabbis understood these two different commandments to represent distinct dimensions of the filial obligation.

> Our rabbis taught: What is reverence (*mora*) and what is honor (*kavod*)? Reverence means that he [the son] must neither stand nor sit in his [the father's] place, nor contravene his words, nor decide an argument in which he is involved. Honor means that he must give him food and drink, clothe and cover him, and lead him in and out. (BT *Kiddushin* 31b)[14]

Reverence, *mora,* is preserving our parents' dignity. This commandment relates to the attitude of respect that is due our parents. The text identifies and prohibits behaviors that might compromise the parent's dignity. Even if our roles have shifted and we are now caring for our parents, we are called to allow them to retain their place. We must not usurp their role or authority.[15] Moreover, we are not to make decisions that fail to respect their wishes.

In contrast to *mora*, which is attitudinal, honor, *kavod*, revolves around providing for our parents' material and concrete needs. This *mitzvah* obligates us to ensure that our parents have adequate shelter, food, clothing, and transportation. It is our responsibility to see that they receive exemplary care. In the face of these overwhelming obligations of *mora* and *kavod*, how are we to balance competing claims? How is the caregiver's quality of life to be factored into the equation?

COUNTERVAILING VALUES

Our sources suggest that in addition to the compelling need to provide for quality of life for an elderly parent, the quality of life of the caregiver is also worthy of attention. We can find examples of adult children making choices to fulfill their own dreams and aspirations as early as Abraham. *Midrash Rabbah* tells us that Abram, as he was then known, left his father, Terah, to follow the divine call. The *midrash* takes pains to explain why the verse reporting Terah's death—"And Terah died in Haran"—appears before God's call to Abram, despite the fact that (as can be inferred from the biblical text) his death actually occurred sixty-five years later. The *midrash* explains that Abram was reluctant to leave when called by God, as he feared people would criticize him, saying, "He abandoned his father in his old age." God reassures Abram, stating, "I exempt you (*lekha)* from the duty of honoring parents, though I exempt no one else from this obligation. Moreover, I will record his death before your departure." (*Genesis Rabbah* 39:7).

This text is provocative. Reading it, we may wonder if this is a parallel to the *Akedah*, in which God's demands of Abraham supercede his obligations to his human family. If so, is it truly only a one-time exemption, or might a contemporary son or daughter have a calling that could be considered adequate justification for putting a parent's needs second? It is worth noting that the *midrash*'s solution to the dilemma actually only removes the *appearance* of a choice to put calling before parent care, as Terah truly did not die until years after Abram's departure.[16]

In a fascinating recognition of the agony of some caregiving situations, Maimonides suggests that an adult child might need to delegate caregiving tasks in certain circumstances:

> If one's father or mother should become mentally disordered, he should try to treat them as their mental state demands, until they are pitied by God [they die]. However, if he cannot endure the situation because of their extreme madness, let him leave and go away, deputing others to care for them properly. (MT, Laws of Rebels 6:10)[17]

Significantly, the criterion for when that point is reached is the adult child's *subjective* experience. Only the adult child can say when he or she has reached his or her limit. Contrary to the first-blush impression we received on looking at BT *Kiddushin* 31a, caregivers are entitled, or perhaps even *must* attend to their own needs and limits.

Traditional sources reflect a recognition that caregivers are often balancing multiple competing caregiving responsibilities. A married woman, for example, is exempt from the obligation to care for her parents, since her obligation to her husband is given primacy.[18] The care given to one's parents apparently must not undermine the well-being of one's spousal relationship. For example, Maimonides rules that a husband may refuse to allow his wife's parents to visit in his home:

> A man who tells his wife, 'I don't want your father and mother, brothers and sisters to come to my home' is to be obeyed. She should visit them in their home monthly and on every holiday, and they should come to her only in unusual circumstances, such as illness or birth, for a man is not to be forced to bring others into his domain. (MT, Laws of Marriage 13:14)

Maimonides also extends this right to the wife vis-à-vis the husband's parents.

> Also if the wife says, 'I don't want your mother and sister to come to my [home], and I will not live in a shared courtyard with them, because they are mean and cause me grief,' she is to be obeyed, for a person is not to be forced to have others live in his domain. (MT, Laws of Marriage 13:15)

Interestingly, while the husband's right seems to be unqualified, in this case, the wife must offer a justification, or, according to some authorities, prove her assertion.[19] While the particular gender distinction here may repel us, the value of preserving the spousal bond is evident.

Finally, in the relationship between adult children and their parents, our tradition offers norms for the parent as well as the child. While the child clearly carries a weighty burden of obligation, the parent is reciprocally bound. For example, the parent must not make things harder upon the child, or be overly demanding.

> Although we are commanded (regarding honoring parents), a person is forbidden to *add to the burden upon his child* and to be particular regarding his honor (*lehakhbid 'ulo al banav u'ledakdek bikhvodo imahem*), lest he cause them to stumble [sin],[20] rather, he should waive [his honor], and ignore [behavior which is not strictly in keeping with the *mitzvah*], for when a parent waives his or her honor, it is [effectively] waived. (MT, Laws of Rebels 6:8)

We have seen that the values embedded in our tradition impel us to provide for our aging parents' physical care, and to maintain their dignity, while also attending to our own personal and familial well-being. We have also seen that parents are obligated to avoid placing their children in impossible binds.

The ethic of care described above urges us to conduct moral reasoning in the context of a particular set of relationships, and to take interdependence and mutuality into account. This approach brings the endeavor of ethics to our concrete realities in a way abstract discussion cannot. We can thus best explore how our tradition's values outlined above illuminate contemporary caregiving dilemmas through the analysis of a narrative of a paradigmatic caregiver-parent crisis.

> Myra and her husband Sam raised their family in Queens, New York. When they retired, in 1983, they moved to Florida, where they bought a small condominium. They made new friends, and reconnected with friends from earlier parts of their lives. Their daughter, Roberta, lived in Boston with her husband, Michael; their elder daughter, Sherry, died several years ago.
>
> After Sam's stroke five years ago, he was able to walk with a walker, but he could no longer drive. Suddenly, life in the condominium was no longer feasible. Roberta offered to help Myra and Sam move closer to her home, but Myra and Sam chose to remain in Florida, where their friends were. Roberta helped Myra and Sam to move into an independent living facility.
>
> Roberta and Michael again offered to help Myra to move North after Sam's death six months later, but she felt strongly that she wanted to stay in Florida. She made new friends and enjoyed the activities in her building. Roberta, Michael, and their two daughters visited two or three times a year.
>
> Four years ago, Roberta noticed that her mother was beginning to be forgetful; in the following year, a neurologist diagnosed Myra with Alzheimer's disease. Roberta, who had recently opened a psychotherapy practice after years of working part-time, began to travel to Florida every two months; on these trips,

she took Myra to the doctor, arranged a companion for a few hours a day, and watched her mother decline. Whenever they discussed the possibility of moving near Roberta, Myra refused. Although many of her friends had died or moved to other facilities, she felt at home in Florida, and didn't want to move to a place where she would know no one.

Although the frequent trips were financially and emotionally draining, Roberta was committed to respecting her mother's wishes. She worried about her mother constantly, and called her every morning and evening. Her mother's confusion was increasing. Finally, Roberta got a call from the manager in her mother's building. Myra had wandered away from the complex, and, in her disorientation, could not find her way home. Roberta would have to move her mother to a nursing home or assisted living facility.

Should Roberta move her now eighty-six-year-old mother to a facility in Boston, or find one in Florida, as her mother wishes? We have seen that there are obligations on *both* sides of the elderly parent–adult child relationship. The parent has the right to make her own decisions, but not to "intensify the child's burden." In an effort to apply our values to this case, let us first examine each party's quality-of-life concerns. In her newly frail state, Myra needs *kavod*, care, more than ever. She cannot arrange for her care, nor assure that it is competently or humanely provided. Beyond *kavod*, Myra needs to have her dignity preserved (*mora*). She needs to have her preferences and her values respected. Staying in her familiar surroundings reduces strain on her, especially as her confusion grows.

Roberta, on the other hand, needs to be able to care for her mother. She is obligated, and truly wishes, to attend to Myra's well-being. At a distance, she is not able to do this to her satisfaction. Even if she were to increase the frequency of her travel, she would not be content with the level of involvement and advocacy she could contribute toward her mother's care. She cannot follow up on medications, appointments, home care, and medical care from a distance of one thousand miles away. In addition to her need to care well for her mother, Roberta needs to fulfill her responsibilities to her husband and her clients. Finally, Roberta needs to stay well, physically and emotionally.

Myra's and Roberta's quality-of-life needs are in painful tension with each other. While Myra was cognitively intact, she could choose to "waive her honor," absolving Roberta of her obligations of *kavod*, or at least limiting her expectations of her daughter with understanding of the limits imposed by her choice to be far away. Roberta might not *feel* absolved, but she would certainly not be accountable for the gap between the kind of care she would ideally like to provide and that which would be possible from a distance.

Now, Myra is confused and not capable of affirmatively waiving the honor due her, and Roberta is unable to retreat from her obligations. Sadly, due to her decision to stay in Florida, Myra has caused Roberta to stumble. Roberta is faced with either stumbling literally by pushing herself to exhaustion in her effort to care well for her long-distance mother, or stumbling in her obligations of *mora* by contravening Myra's stated wishes and moving her close by.

Roberta is now bearing not only the burden of her own multiple obligations, but also the burden of the choices her mother has made. This unreasonable burden tips the scales in the quality-of-life equation, and justifies Roberta's decision to move her mother to an assisted-living facility near her home in Boston. In taking this action to ensure that she can provide *kavod* for her mother, Roberta should endeavor to foster her mother's dignity throughout the process. She should, if possible, involve her mother in the choice of the facility, and in furnishing her room, as well as in the process of discarding belongings and packing up her Florida apartment. Roberta might create a ritual of leave-taking, so that her mother can bid farewell to the friends, surroundings, and memories in that home she has loved so much. Involving Myra in the transition, heeding her wishes within the confines of necessity, and honoring the pain she is feeling will allow Roberta to relate to Myra as a *subject* in her own life, not an *object* of care. Doing this will enable Roberta to continue to fulfill her obligation of *mora*.

> Rav Assi had an aged mother. "I want jewels," she said, and he got her jewels. "I want a man," she said, and he said, "I will look for one for you." [When she said], "I want a man who is handsome like you," he left and went to the Land of Israel [from Babylonia, where they lived]. (BT *Kiddushin* 31a)

While Rav Assi was an exemplar of extreme filial piety, he, too, had limits. At a certain point, he had to give precedence to his own quality of life and well-being and literally distance himself from his mother. Dedicated son that he was, we would imagine that he arranged for others to provide the care he was no longer able to render personally.[21] In the case of Roberta, another exemplary caregiver, giving precedence to her own quality of life requires bringing her mother closer to her, but the principle is the same in both cases.

A CONCLUDING HOPE

R. Shimon b. Yohai said "[T]he most difficult of all *mitzvot* is 'Honor your father and your mother.'" (*Tanhuma Ekev* 2). Nothing can take away the complexity, intensity, and weight of caring for those who brought us into the world. On the other hand, there is nothing in our tradition that says that we must be consumed by caregiving, wrung dry, left with nothing for ourselves and our own families. Caring for elders in our families demands more resources than any single caregiver can muster alone. Just as we have learned that it takes a village to raise a child, so too, may we come to realize that it takes an entire family, and, actually, an entire community to foster quality of life for frail elders and their caregivers.

REFERENCES

Dychtwald, K. (1990), *Age Wave: How the Most Important Trend of Our Time Will Change Your Future*, New York: Bantam Books.

Gilligan, C. (1982), *In a Different Voice: Psychological Theory and Women's Development*, Cambridge: Harvard University Press.

——— (1987), "Moral Orientation and Moral Development," in *Women and Moral Theory*, ed. E. F. Kittay & D. T. Meyers, Rowman & Littlefield, 19–33.

Gordis, D. H. (1989), "Wanted—The Ethical in Jewish Bioethics," *Judaism* 38, 28–40.

Groenhout, R. (2003), *Theological Echoes in an Ethic of Care*, Erasmus Institute, University of Notre Dame.

Noddings, N. (1984), *Caring: A Feminine Approach to Ethics and Moral Education*, Berkeley: University of California Press.

Poirier, S. and Ayres, L. (2002), *Stories of Family Caregiving*, Indianapolis: Center Nursing Publishing.

Rieger, Miriam (2004), *The American Jewish Elderly*, United Jewish Communities Report Series on the National Jewish Population Survey, 2000–2001.

Teutsch, D. A. (2001), "Values-Based Decision Making," *The Reconstructionist* 65, no. 2, 22–28.

Tronto, J. C. (1994), "Beyond Gender Difference to a Theory of Care," in *Feminism*, Vol. II, ed. S. Moller Okin and J. Mansbridge, Brookfield, VT: Edward Elgar Publishing Company, 318–37.

NOTES

1. This term was coined by Dychtwald, K. (1990).

2. The National Jewish Population Study (Rieger, 2004) found that the proportion of Jews who are elderly is more than twice the average proportion of elderly worldwide.

3. The National Jewish Population Study (Rieger, 2004) reported the lower rates of fertility of Jewish women than their non-Jewish peers over the past generation, as well as significant geographic mobility—35 percent of Jewish adults lived in a different location than five years before, 10 percent in different cities, 10 percent in different states, and 2 percent in a different country.

4. See the essay by Deena Zimmerman in this collection, chapter 7.

5. See, for example, Gilligan (1982, 28–32).

6. See Noddings (1984).

7. Gilligan (1987, 24) suggests the importance of interdependence and reciprocity in the ethic of care.

8. For a lucid summary of the various positions in the debate between the ethics of care and the ethics of justice, see Poirier and Ayres (2002).

9. A fascinating model for infusing the theological dialogue with the gleanings of care theory is offered by Groenhout (2003). Groenhout investigates elements of the worldview of the ethic of care that are congruent with Jewish and Christian ethics, and those that might offer correctives to significant "blind spots." Importantly, Groenhout suggests that the ethics of care's focus on human finitude and interdependence can enrich theological approaches to ethics.

10. See Gordis (1989, 29).

11. See Teutsch (2001).

12. Although it is beyond the scope of this chapter, it would be worthwhile to investigate filial caregiving in Jewish sources outside of the rabbinic realm, such as folktales and folk songs, and modern Yiddish and Hebrew literature.

13. The literal translation of the verse, "A man shall revere his mother and his father," has been rendered here in a more inclusive manner.

14. The text's references to both caregiver and care receiver are male. We cannot assume that daughters and others are excluded, but we might wonder what the rabbis would have said if describing the obligations of *mora* and *kavod* for a daughter toward a father, or a son toward a mother. Maimonides suggests that both sons and daughters are obligated, though the wife's capacity to fulfill the obligations is limited by her other responsibilities, presumably to her husband and children. (MT, Laws of Rebels 6:6)

15. Interestingly, some commentators read "and shall not decide an argument" to mean that the son should not play the role of decisor, even on behalf of the father's position, for to do so would be to assume a superior role to the father's. Cf. SA, YD 240:2.

16. The text offers another justification for this abandonment: Terah is wicked, and thus was in the category of those who are "called dead even while they are alive."

17. Rabad disagrees, suggesting that if the child will not care for the parent, no one else will, either. David ben Zimra (Radbaz, sixteenth century) defends Maimonides's position, suggesting that there are times that non-relatives may have an easier time dealing with the elder's behaviors than an adult child, who is emotionally involved.

18. Applying such an exemption of women would be greatly at odds with contemporary gender roles, wherein, most commonly, women assume caregiving obligations not only for their partners and children, but for their own and their spouse's parents. Nonetheless, this exemption hints at the challenges of juggling multiple caregiving responsibilities for intimate others.

19. Magid Mishneh (citing a responsum of Alfasi) suggests she must prove in court that they are causing her suffering and/or causing strife in her marital bond. She is required to do this, since the husband is presumed to own the house and thus has complete authority to determine who visits. In this reading, the wife's right is only to be free from suffering in the home she shares with her husband.

20. The reference is to Leviticus 19:14, "Do not place a stumbling block before the blind." This term is used to refer to causing another to sin (stumbling in the most consequential sense).

21. Interestingly, the Talmud goes on to report that after settling in the Land of Israel, Rav Assi learned that his mother was on her way to him. He asked for advice from his teachers about whether he could leave the Holy Land to go to meet her; he did eventually set out toward her, but learned that she had died before he got to her. I have wondered in reading this text if he was actually seeking to meet his mother or escape her.

7

Family Obligations and Caregivers' Quality of Life: Thoughts from Halakhic Sources

Deena Zimmerman

Our purpose in this discourse is to broaden and deepen bioethical conversation in Jewish life by bringing together people from different experiences, backgrounds, and training. I bring to the questions before us a dual perspective: that of a practicing MD and that of someone who deals on a daily basis with the interaction of *halakhah* and medicine.

What may a Jewish perspective contribute to answering the question, "How should the quality of life for the patient be weighed against the quality of life of volunteer caregivers and their families?" My response is first to note that Judaism has a centuries-old legal tradition that can be looked upon for guidance in dealing with modern dilemmas. While its attitudes may not always be those with which we are familiar, and some cases may in fact make us a bit uncomfortable, it is just such a fresh look that may sometimes undermine our preconceived notions and thus help us grapple with difficult situations. To demonstrate this, I want to review a number of topics in *halakhah* that touch on the questions asked. My goal is not that my words might provide final answers to specific situations—since *halakhah* is case based—but rather to present principles that can be worked with.

The first principle that is apparent in the halakhic sources is that the role of caretaker comprises two obligations, the physical and the fiscal. The second is that not all caretakers are created equal (Rabbi Nebenzal, 1983,

57–65). The obligation may be different depending on the relationship to the one being cared for. This last statement is not egalitarian, and thus appears to be in clear opposition to American values, but it is what seems to be borne out by the *Shulhan Arukh*, and despite the dangers I am going to forge ahead and present it.

The two points I mentioned, fiscal versus physical and varying obligation based on role, are clearly demonstrated in the case of a husband and wife. The standard *ketubbah* (marriage contract) includes two sets of obligations: of husband to wife, and of wife to husband. According to the Talmud (BT *Ketubbot* 47b), these should be regarded as paired specific exchanges, and this legal structure has been preserved in the codes of Jewish law. The sources relate to the obligations of the husband to the wife as protection for the wife. The obligations of the wife to the husband are explained as a method of making the exchange more even, so as not to deter men from marriage. It is important to know that even at the time of the Talmud, if a woman had the means to support herself, the couple could negotiate changes from the standard assumptions.

For our purposes we will focus on the standard arrangement as recorded in *Shulhan Arukh, Even Ha-Ezer* 79. There it is stated that the husband is obligated to provide for his wife's *mezonot*, sustenance, in return for the products or income accrued from *ma'ase yadeha*, her household and professional activities. Included in sustenance is an obligation to cover medical expenses. Medical expenses here mean anything that she needs for her recovery or care and would thus include healthcare attendants in addition to medications. He is not obligated to personally undertake her care, but is obligated to assume the financial responsibility. In the event of his death, this financial responsibility is passed on to his sons, even sons from a previous marriage. The wife, on the other hand, does not have this financial obligation to her husband. It is however presumed that she will provide physical caretaking, cooking, cleaning, serving food, in sickness as well as in health. While at the onset the couple can negotiate other arrangements, the de facto plan shows what is assumed to be the most common arrangement.

This standard arrangement does not mean that a husband cannot care for his wife or that a wife cannot pay for her husband's expenses. Ideally this mutual caring will take place automatically. However, if one member of the couple is unable or unwilling, it is important to have a clear outline of what is required.

The law recorded in *Shulhan Arukh* seems to directly address our issue. As mentioned, a husband has to pay for his wife's medical care. Nevertheless,

> [i]f the husband sees that his wife's illness is prolonged, he can demand that she pay [for her healing] out of the money he owes her as part of the marriage agreement [*ketubbah*]. However, it is not proper to do so. (SA, EH 79:3)[1]

This source juggles between the husband's strict financial obligation and an injunction to go beyond this for the good of the care recipient, his wife.

The differentiation between a physical and financial responsibility for providing care also appears with respect to the relationship of a child and parent. The rules of the obligation of children to their parents are outlined in the *halakhot* of *kibbud av va-em* and the *halakhot* of a father's obligation to his children. The former is usually translated—following the source in the Ten Commandments (Exodus 20:12)—as "honoring one's father and mother." However, as specified in the *Shulhan Arukh*, these obligations define much more of a caretaking role. The following is from *Shulhan Arukh Yoreh De'ah*, which will be cited henceforward as SA YD:

> What is "honor"? Providing food and drink, clothing and covering, taking in, around and about. (SA YD 240:4)

In this caretaking role of *kibbud*, the primary obligation of the child to his or her parents is a physical one. As outlined in the next clause, "The child is obligated to provide these services, as a personal obligation" (SA YD 240:5).

It appears that the preferred method would be to do these actions oneself. However, if this is not possible due to distance, for example, the child can arrange to make sure it is done. However, what is required is the personal effort to make the arrangements, not merely to subsidize the financial cost. In fact, the cost of the care of the parent is paid first from the parent's assets. Only if the parent does not have assets and the child does would a court dun the cost from the child. As we shall see later, this cost can be charged to the child's *tzedakah* account. If the child does not have the money, he is not obligated to beg to get it. If there are a number of children, then the burden is split in accordance with each one's ability to pay.

If is of interest that the halakhic tradition places great emphasis not simply on the help and services themselves, but on the accompanying interpersonal attitude:

> He should give with a pleasant countenance. Even if he provides five-star fare but does so with an angry face, he is punished. And so the opposite, even if he requires his father to work at the grindstone but pleasantly explains and demonstrates that it is for the father's own good, the child merits the world to come. (SA YD 240:4; my English paraphrase.)

In the laws of the *kibbud av va'em* we see the problems of competing loyalties. *Kibbud av va'em* is an obligation that falls on both men and women. A married woman, however, is exempt from her ability to perform the commandment. This does not preclude her from doing so if this generates no family tension, but in the case where the husband objects, her primary obligation is to her spouse, not her parent:

> Both a man and a woman are equal in the obligation of honor and fear of father and mother. However, a married woman cannot do this, as her primary obligation is to her husband; therefore, she is exempt from the honor of her father and mother while she is married. If she becomes divorced or widowed, she is obligated. (SA YD 240:17)

This clearly shows that the married woman's exemption is not due to gender but to life situation. Prior to marriage, she was obligated. If divorced or widowed, she is again obligated. This also does not mean that her parents have no rights but rather that there is a shift of responsibility. As stated in SA YD 240:24, "A man is obligated in the *kibbud* of his in-laws." For the wife, however, any obligation to care for her in-laws is subject to her primary obligation toward her husband.[2]

I am fully cognizant of the fact that this non-egalitarian approach may be disturbing to many. I am not sure that I myself am totally happy with it. However, if we look at what happens in real life, we see that studies show that in the majority of cases it is the woman who takes on the primary caretaking burden even if she has her own professional career (Navaie-Waliser, Spriggs, and Feldman 2002). Research also shows that this role can be detrimental to the woman's health and well-being (Navaie-Waliser et al., 2002; Mui, 1995). Therefore, one should think about whether women caretakers are better off in a supposedly egalitarian modern society or under the halakhic system that indicates that care of parents is not just a woman's role so that in fact a woman is somewhat protected.

When thinking about the burdens of caretakers, a scenario that often leaps to mind is that of a parent with failing mental ability. We sometimes think of this as a modern problem that is a result of a longer lifespan. However, just such a scenario is laid out in the medieval Jewish codes:

> One whose father or mother has become demented should try to care for them while accommodating their mental condition. However, if he is unable to handle this, as they have become exceedingly incompetent, he should depart and leave them and instruct another to appropriately care for them. (SA YD 240:10)

This teaching provides an initial answer to the question posed at the outset. At a certain point, the child's quality of life—his or her ability to cope—takes precedence over the needs of the parent, even though these are presumably better served by the child directly than by proxy. While it can be argued that hiring a professional cook, for example, may be better than amateur cooking on the part of the child, my personal experience as a doctor shows a clear benefit of the on-site supervision of a family member. Nevertheless, if this is too great a strain on the child due to the parent's dementia, he is permitted to forgo direct personal involvement.

In the case just cited, the adverse effect seems to be on the caregiver's quality of life. In other cases, where caregiving children or a spouse are themselves in poor health, the caregiving role may endanger their health. This kind of conflict is more often discussed when dealing with professional caregivers—doctors, nurses, paramedics—but it can be no less real in the family setting as well. Here we can look to the well-known talmudic instruction regarding two travelers, lost in the desert with a single bottle of water, sufficient for only one to survive (BT *Bava Metsia*, 62a). Just as the person who holds the water takes precedence, so one does not have to heal others at the expense of one's own life. Similarly, one is not *obligated* to donate an organ while alive (although this would be *permitted*, provided that the donor will survive). Thus if the caregiver would be endangered by the physical toll of caring for another, this might exempt her from the obligation.

The above sources all refer to the care of ill adults. It is harder to come up with halakhic sources related to the care of ill children. However, from my experience as a pediatrician I can say that balancing the needs of the caregiver and the care recipient in such a scenario also needs to be addressed.

The obligation to care for even a healthy child is on the father—once again, non-egalitarian but a bit eye-opening. The primary role seems to be to get the child started on an independent life. For a boy this includes teaching him a trade, finding him a spouse. For a girl, it means providing for her needs until maturity. When dealing with a child who will not be independent due to mental or physical incapacity, the best precedents for these cases appear in the laws of *tzedakah*. This word is generally translated as "charity," which suggests voluntary giving, whereas *tzedakah* is an obligation. As spelled out in SA YD section 149, a certain portion of one's income must be set aside for this purpose. There is also an absolute minimum of "one-third of a *shekel*" which even the poorest of the poor is obligated to give, since his own poverty does not exempt him from this commandment. The generally recommended share is 10 percent, with a maximum of 20 percent of one's income. In the (pre-emancipation) autonomous Jewish community, people not giving their allotted amount could be compelled to do so.

The individual has discretion as to how to apportion the money that he or she gives as *tzedakah*. It is very clear, however, that charity begins at home. The halakhic codes list close relatives as having priority in the distribution of individually given *tzedakah*. With regard to supporting children, for example, the background assumption is that a man's specific legal obligation to support his young children extends up to the age of six. Beyond this, his obligation to them falls within the ambit of his *tzedakah* duties. Such giving, and likewise giving to other family members and relatives, takes precedence over addressing the needs of other claimants (SA YD 251:3).

If we seek to focus on the burdens and issues faced by families with children needing long-term care, it is not easy to find express relevant teachings. It is worth noting that in bioethical discourse in general, these issues have not been addressed as much as those of families dealing with elderly relatives.

The tradition does seem to reflect some acceptance for the care of those with disabilities in the community. This might be deduced from the fact that in halakhic sources we find multiple references to the *heresh* (the deaf mute), the *shoteh* (the mentally incompetent), and the *suma* (the blind). These include, for example, discussions of their capacity to conduct business and to enter into marriage. Perhaps, then (although there is no firm proof of this), it is a Jewish value to encourage caring for those with disabilities in the community. Acceptance of those who are different is beautifully illustrated by the existence of a blessing that is to be recited upon encountering those who are different.

> Blessed are thou, LORD our G-d, King of the Universe, who has made different creations. (My translation, DZ.)

As evidence has shown that children with certain conditions raised at home can accomplish so much more, the trend has been to encourage care at home. However, this often takes a tremendous toll on the family (Thyen et al. 1999; Kirk 1998). A personal account of one family's struggle and their eventual decision to go against the trend is found in the book *Before and After Zachariah* by Fern Kupfer (Academy Chicago Publishing, April 1988).

As part of my research for this paper I interviewed a neighbor who is the mother of five children, three of whom suffer from familial dysautonomia or Riley Day Syndrome. This genetic disease, more common among Ashkenazi Jews, leads to severe physical disabilities due to scoliosis, multiple hospitalizations due to instability of the autonomic nervous system, and variable degrees of developmental delay. The oldest two now run our supermarket as well as a business in the Israeli staple of sunflower seeds. The second attended regular school through high school. Her youngest was about to start first grade when she developed encephalitis of an unclear etiology from which she has not recovered. After one year without progress, she spent two years in a chronic-care facility, and then passed away.

I asked her about her concerns, as one who has chosen to raise her children at home and provide for them maximal opportunities, and she then expressed two major concerns. Perhaps surprisingly, the number one concern of this mother is the lack of understanding and sensitivity on the part of some members of our community. This is shown, for example, in the practice of talking about the affected children in their hearing as if they cannot understand, whereas they certainly do. I raise this here in passing, to call attention to a matter that should be addressed within our discussion of a Jewish perspective on ethics. Her second major concern—which is directly related to the question at hand—will be reported below.

Unlike the situation in caring for a spouse or parent where it is likely that the caregiver will outlive the one needing care, with children this may very well be the opposite. When the ill child outlives his parents, the burden of care may fall on his or her siblings. This last situation is a much under-discussed topic in bioethics. A recent book has been written by psychologist Jeanner Frazer, titled *The Normal One* (Simon and Schuster, 2002). It provides a much-needed description of the problem, but less in the way of a solution. What struck me, however, was the bitterness with which the author, a psychologist, writes about her experience and those of other siblings.

This demonstrates the seriousness of the issue, which was in fact the second major concern expressed by my neighbor. Parents have the right to distribute their resources (while they are alive) among their children, although we have already seen from the biblical story of Joseph and his brothers the family strife that can cause. For the role of brother (or sister) caring for brother (or sister), I have not been able to find a precedent in the traditional sources. Yes, supporting an ill sibling would also count as *tzedakah*; but more than that I cannot say. Once again, further exploration within a Jewish perspective is needed.

In summary, I think the Jewish tradition does not provide clear-cut answers. It does, however, lay out the dilemma in a unique way, including the legitimacy of considering the caretaker. This dilemma is summed up in the "Ethics of the Fathers" (*Pirke Avot*, 1:14): "If I am not for myself, who is? But if I am only for myself, what am I?" Within Jewish tradition it is legitimate to weigh the competing needs of the caregiver and the one in need of care. The proper balance needs to be determined on a case-by-case basis.

REFERENCES

Kirk, S. (1998), "Families' experiences of caring at home for a technology-dependent child: a review of the literature." *Child: Care, Health and Development* 24, 101–14.

Mui, Ada C. (1995), "Caring for frail elderly parents: a comparison of adult sons and daughters." *Gerontologist* 35 (1), 86–93.

Navaie-Waliser, M., Spriggs, A., and Feldman, P. H. (2002), "Informal Caregiving: Differential Experiences by Gender." *Medical Care* 40 (12), 1249–59.

Navaie-Waliser, M., Feldman, P. H., Gould, D. Z., Levine, C., Kueerbis, A. N., and Donelan, K. (2002), "When the Caregiver Needs Care: The Plight of Vulnerable Caregivers." *American Journal of Public Health* 92, 409–13.

Nebenzal, Avigdor (1983), *Hazikna Ba-Yahadut*. Jerusalem: Machon Schlesinger.

Thyen, U., Kuhlthau, K., and Perrin, J. M. (1999), "Employment, Child Care and Mental Health of Mothers Caring for Children Assisted by Technology." *Pediatrics*, 103 (6, Part 1), 1235–42.

NOTES

1. The English version here, and in all quotes from the *Shulhan Arukh* in this essay, are my own. I did not always strive for an exact word-by-word translation, but have provided instead a paraphrase intended to convey the law with fidelity but also optimal clarity to a contemporary readership.

2. This is explicated in the comment to this clause by R. Hayyim Yosef David Azulai (HIDA, Jerusalem, eighteenth century), *Birkey Yosef* ad loc.

III

QUALITY OF LIFE AS A GOAL OF MEDICINE: FROM THERAPY TO ENHANCEMENT

8

Reflections on Enhancement, Authenticity, and Aging

*Thomas R. Cole and Robin Solomon**

Studying the "Treatment vs. Enhancement" debate in contemporary medicine is a daunting task. One quickly realizes that there is no clear boundary between the two, and that the term "enhancement" is itself controversial. A wide range of issues are in play—including the goals of medicine, socially just applications of technology, and visions of the good life. Our chapter does not attempt a comprehensive analysis of these issues. Rather, we conceptualize and interpret the debate from different angles, and we then extend the interpretation to anti-aging enhancements, seen from both secular and Jewish perspectives. We reflect on five major themes.

First, we offer a conceptual map of the major issues surrounding enhancement and provide a historical context for understanding its blurred boundaries. Next, we examine Carl Elliott's *Better Than Well* (2003), a book that brilliantly interprets enhancement technologies in American culture. Third, we argue—against Elliott—that a robust concept of authenticity and careful, pluralist religious scrutiny of enhancement technologies can provide considerable moral guidance. Fourth, we look specifically at "anti-aging" medicine and research as a brief case study of enhancement technologies.

---

*Robin Solomon did most of the research and writing in section 1. Thomas Cole did the primary work in sections 2 through 5.

Finally, we ask the question, "How should Jews think about anti-aging enhancements?"

1—MAPPING THE THERAPY/ENHANCEMENT DEBATE

The accelerating pace of medical progress today is dizzying. Mapping of the human genome, new technologies of genetic manipulation, the rise of molecular medicine and nanotechnology—these advances and others promise new, more effective measures for the prevention and treatment of disease. While new possibilities give rise to new (sometimes unrealistic) hopes, they also generate (or push us to reformulate) fundamental questions about the goals of medicine and the use of new technologies. Medicine increasingly has the capacity not just to turn pathology into normality, but also to "improve" on normality—to make people more (conventionally) beautiful, to increase their height, perhaps even their IQ or other socially desirable characteristics. It may soon be possible to retard or reverse biological aging.

Such current treatments and future possibilities have given rise to the "Therapy vs. Enhancement" debate. One can argue that treatment of nonpathological conditions to gain improvement is not in fact therapy, but rather constitutes enhancement and therefore has a different (and unsettled) moral status. Is enhancement a proper use of medical resources? Is it permissible? Obligatory? If we answer "yes" to any of these questions, then who ought to benefit, what are the moral limits of enhancement, and who should pay the bills?

The "Therapy vs. Enhancement" debate is so contentious and interesting because the two concepts at stake are no longer clear. Is there a valid distinction between them? And if so, where should the line be drawn between treatment and enhancement? Let us trace some of the arguments for and against the use of this distinction.

In genetic medicine, the distinction is often used in order to argue that only therapeutic uses of genetic engineering fall within traditional medical boundaries. From this perspective, enhancement uses are not seen as addressing a legitimate medical need and hence lie outside medicine's proper domain (Juengst, 1997).

Opponents of the distinction claim that "enhancement" in this context cannot be defined in a way that will convincingly distinguish it from therapy. Consider the following much-rehearsed example of two ten-year-old children, both in the third percentile of their age group. One has a diagnosed deficiency of human growth hormone (hGH) and the other does not. Using the traditional medical criteria, only the diagnosed child would receive biosynthetic hGH: for him this is deemed "therapy," whereas for the other the same treatment is defined as "enhancement." Assuming that without treatment both children would achieve the same adult stature and that no other diseases are associated with the hGH deficiency diagnosis, some argue that in this case the distinction reflects no valid medical consideration. The real "problem" is that parents do not want their children to suffer the social disadvantages of being short. Hence the decision of which child to treat should not rest on the traditional medical criteria of diagnosed pathology.

Commentators Gerald McKenny and Estuardo Aguilar-Cordova (1999), and David Resnik and Pamela Langer (2001) advocate more neutral terms, such as modification or manipulation, especially for genetic constructs, again because they see no useful defining boundary between what a treatment or what an enhancement would be. Rather than rely on the treatment/enhancement distinction, they suggest prioritizing possible interventions and focusing on the risks and benefits of particular kinds of interventions.

Christopher Newell (1999), and Jackie Leach Scully and Christoph Rehamann-Sutter (2001) are among those who argue that the distinction is arbitrary, useless, or even damaging. They worry that if such a distinction is built into regulations, normative standards for human identity and embodiment will produce discrimination against people with disabilities. Interestingly, a recent survey of genetic scientists reported that while they were overwhelmingly against using technology for enhancement, they acknowledged that they could not adequately distinguish between a "disease"—which it would be appropriate to treat—and a condition that does not warrant enhancement (Rabino, 2003).

Yet many commentators are able to offer definitions of the distinction. Glenn McGee (2000) defines enhancement as that which can "offer changes in human capacity." Norman Daniels (2000) argues that enhancement consists of interventions that produce biological changes above and beyond treatment of disease and the restoration of species-typical normal functioning. Eric Juengst (1997) argues that a distinct line can be drawn if we accept certain health problems as distinct disease entities and if we accept that disease prevention is a legitimate goal of medicine.

Thomas Murray and Norman Daniels are among those who advocate a limited use of the distinction. They see it as a good starting point for discussion, but argue that we should not expect it to provide much normative guidance. Similarly, Eric Parens (2002, May 13–14, 154) suggests that the distinction can help us to think about what universal health care should include and can be used as a tool to "begin critiquing some social practices." Parens notes that "if say, shyness isn't a disease, perhaps medicine ought not to treat it and medicalize it" (Parens, 154). The distinction can also help us affirm the diversity of human beings. Parens acknowledges, however, that it will not help us with the "schmocter" problem—people with access to new technologies willing to sell them to anyone for enhancement purposes.

Arguments for and against the use of medicine for enhancement must take account of the wide variety of enhancements, which fall roughly into three categories: cosmetic, pharmacological, and genetic. Opponents of genetic enhancement, for example, claim that it amounts to "playing God," or that it violates human nature and defiles human dignity (Fukuyama, 2002; Kass, 2002). Franklin Miller, Howard Brody, and Kevin Chung (2000) argue that enhancement is not a legitimate goal of medicine and we do not have a duty to secure everyone's total well-being or help them attain a selfish form of "authenticity." There are also strong social justice concerns against enhancements that confer a competitive advantage on those treated. If only a few get enhancement, then we are promoting inequality. If everyone gets enhancement, then we are promoting a zero-sum situation in which no one benefits except the owners of the technology.

How is enhancement supported? Libertarians hold that each individual should be free to pursue or use enhancement technology for themselves, their children, or their future children. Paul Root Wolpe (2002) suggests human beings might use enhancement because it is natural to push evolution forward. There are the status quo supporters who say that since we already let people pursue other avenues of enhancement, use of newer technology is no different. Art Caplan (2002) uses the example of parents paying for the Kaplan test prep to get their students better scores on the SATs. He claims that there is no ethical difference between this and pursuing genetic enhancement.

There are also social justice arguments supporting enhancement in order to better the lives of the less-well-off, thus balancing the natural lottery. Another position supports enhancement technologies because they can help people become their "authentic" selves. Ironically, such pursuit of authenticity carries the risk of inauthenticity. Enhancement technologies can lead to the possibility of not leading one's own life. In searching for something more, we may be turning away from our uniqueness, from finitude, suffering, and vulnerability that make us human (Parens, 1998).

Extracting from the complicated scholarly literature, there seem to be three basic positions regarding the issues at stake in the argument over the treatment/enhancement distinction:

(1) Create a clear distinction between accepted treatment and unacceptable enhancement using the traditional, limited view of the goals of medicine.
(2) Weigh the risks and benefits of modifications on a case-by-case basis.
(3) Allow enhancement based on market considerations or on the grounds of egalitarianism.

Families, physicians, and policy-makers will be debating and seeking to operationalize these positions for the foreseeable future.

2—THE HISTORICAL AND CULTURAL CONTEXTS OF THE THERAPY/ENHANCEMENT DEBATE

Contemporary confusion about the distinction between medical treatment and medical enhancement exists in a larger historical context. Sometime around the last quarter of the twentieth century, the wave of postmodernism reached American shores (Best and Kellner, 1991; Harvey, 1989; Rosenau, 1992). It is important to remember that the term "postmodern" refers not only to a range of cultural and intellectual perspectives but also to a temporal watershed marking a new historical era. Observers like Anthony Giddens use the label "late modern" rather than "postmodern," but no serious observer of contemporary culture doubts that the world has passed into a qualitatively new period of historical time.

Think of the forces at play: the computer and the digital revolution, which created an explosion of information and the speedup of almost everything including the production of new scientific knowledge; the saturation of the self with images generated by all kinds of electronic media spurred by consumer culture; globalization, identity confusion, intensified status anxiety, and the rapid growth of immigration from Asia, the Middle East, Latin America, and, to a lesser extent, Africa. These forces burst old moral, intellectual, religious, and cultural boundaries; they have placed us in a period of the most extensive, frightening, and creative confusion since the Renaissance.

Our era has eroded if not dissolved the thick black lines that had divided modern Western culture (since Descartes and the Scientific Revolution) into self-contained boxes of life and thought. Old dualisms have dissolved into blurred categories and genres (Geertz, 1980): normal and pathological; subject and object; self and other; life and death—to name but a few. It may well be that the only dualisms still standing from the 1950s are AM & FM, and Ozzie and Harriet! Even analogue and digital may soon be swept into the dustbin of history. In terms of Jewish life, think of the rise of Jewish Renewal and the intermingling of Reform, Conservative, Reconstructionist, and even Orthodox ideas and practices.

In the academy, the stability of old disciplinary boundaries has been challenged by widespread border crossings and the rise of interdisciplinary fields in the sciences and humanities. Bioethics itself is a case in point, despite the futile efforts of some analytic philosophers to preserve it as the province of pure reason. One has only to think of the influences of feminism, social sciences, empirical bioethics, narrative ethics, hermeneutic ethics, casuistry, theological ethics, professional ethics—and more to the point here, Jewish ethics—to see the erosion of boundaries.

Some observers fear that postmodern fragmentation of the self, of the literary canon, and of grand narratives entails a dangerous skepticism about knowledge, moral relativism, and a slide into narcissism and decadence.

Others celebrate the breakup of old disciplinary boundaries; they await the development of new unifying theories and point to the emerging reorganization of knowledge into larger interdisciplinary fields such as anatomy and neurosciences, integrative biology, integrative medicine, cultural studies, science and technology studies, and the medical humanities. Certainly, one salutary consequence of the waning of rationalist abstractions is that we have become more sensitive to the embodied, contextual, and relational character of human knowledge and experience.

Medicine, of course, has been plagued by moral and intellectual confusion since at least the 1960s (as described in David Rothman's *Strangers at the Bedside* [1991] and Albert Jonsen's *The Birth of Bioethics* [1998]). Technological innovations by themselves were not the exclusive cause of this, as technological innovation in the twentieth century took place amidst moral, spiritual, and intellectual problems that surfaced in the wake of the collapse of an older, religiously based consensus about the meaning and purpose of human life (Rieff, 1966). It is no wonder that we can no longer take for granted the difference between treatment and enhancement.

3—EXAMINING THE "BETTER THAN WELL" MENTALITY

Even though the preceding sections have shown how difficult it is to draw a line that excludes "enhancement," many continue to try. A cultural analysis is offered by Carl Elliott, who seeks to explain both why enhancement is so appealing and why many of us nevertheless feel deep misgivings about it. After recapping his analysis—which amounts to poignant social criticism—we shall offer our own response.

Carl Elliott's recent book *Better Than Well: American Medicine Meets the American Dream* (2003) looks at enhancement technologies to help understand identity in the contemporary American life. That is, Elliott uses enhancement technologies as a mirror in which we may catch glimpses of our anxious, protean selves. "The issues at stake in medical debates over enhancement technologies are important," he writes, "mainly because of what they can tell us about pathologies in the way we live. The uneasiness that many of us feel about enhancement technologies can tell us something important about selfhood, authenticity, and the good life."

Elliott does not advocate for or against enhancement technologies. Rather, he offers a cultural interpretation of their appeal and broadly assesses the moral benefits and costs of their use. Elliott writes in the tradition of David Reisman, Philip Rieff, and Christopher Lasch—cultural analysts and critics trying to understand character and culture in different historical periods. David Riesman's *The Lonely Crowd* (1950), for example, described the emergence of the "other-directed man" who worked in large bureaucratic settings and tended to lose touch with an inner moral compass. Likewise, Philip Rieff's *The Triumph of the Therapeutic* (1966) described "therapeutic man," who sought merely to feel good, thereby attenuating his moral and communal commitments. Christopher Lasch's *The Culture of Narcissism* (1978) described the "narcissist" individual, unable to internalize moral values or develop lasting relationships.

Like his predecessors, Elliott constructs an "ideal type" of personality to convey a central tendency in society and culture. Elliott realizes that enhancement technologies (of which he covers a very wide range: accent reduction, trans-sexual surgery, cosmetic surgery, anti-depressants, anti-aging) are usually marketed and sold by taking advantage of a person's perception that she is deficient in some way. This has led some observers to argue that the moral importance of enhancement technologies derives from "old fashioned American-style self-improvement." But Elliott argues that identity is a better framework than self-improvement for understanding the appeal of enhancement technologies. In his eyes, the significant question is not so much the social and cultural costs of the "quest to become BETTER, but whether there is any moral cost to becoming DIFFERENT. . . . because much of our ambivalence is . . . about 'what kinds of people we want to be' . . . we have mixed feelings about the visions of the good life these technologies serve" (27).

Elliott's cultural interpretation draws on two master texts: Charles Taylor's *Sources of the Self* (1989) and Alexis de Tocqueville's *Democracy in America* (1840, 1945). From each text, Elliott takes a pillar for his argument. From de Tocqueville, Elliott borrows (as have countless commentators before him) the observation that Americans are chronically anxious about their status because they live in a fluid, competitive market society where identity is rarely fixed but rather is confirmed by the social recognition of others. From Taylor, Elliott borrows the pervasive ideal of "authenticity," which has become a central element of modern identity. Authenticity is the ideal according to which each individual has a unique way of being human and each individual is obliged to live out that uniqueness.

Elliott finds authenticity a morally dangerous ideal, in which being in touch with your feelings has come to be an end in itself. Accelerated (I would say co-opted) by the marketing engine of consumer culture, authenticity in many circles is considered a higher form of life because "it is a life of fulfillment, a life in which you know who you are and live out your sense of your self." Elliott cites Jan Morris' *Conundrum*, a memoir about trans-sexual surgery, as an example of medical treatment motivated by the quest for identity, for the true self waiting to be expressed or found.

Elliott believes that Americans are drawn to enhancement technologies for highly problematic reasons. First, we are driven to alleviate the status anxiety created by marketplace competition and consumer culture. And second, we are addicted to an ideal of authenticity that leaves us with no moral horizons broader than individual self-fulfillment. Elliott's tone is much lighter, ironic, and humorous than critics like Philip Rieff or Christopher Lasch. But his implication is clear: we cannot adequately assess enhancement technologies apart from understanding the cultural pathologies that partly give rise to them.

Better Than Well offers a shrewd, scientifically and philosophically sophisticated portrait of enhancement as a symptom of America's moral and spiritual inadequacy. We are sympathetic to this critique of enhancement in American culture. But key nuances in Taylor and key observations in de Tocqueville are missing from Elliott's account.

For Taylor, the moral value of authenticity does not lie in solipsistic efforts to live out the uniqueness of one's true self. In both *Sources of the Self* and *Ethics of Authenticity*, Taylor makes clear that authenticity worthy of the name contains external social and moral criteria, against which the authentic self wrestles. The resulting personal reappropriation contains external norms to which the authentic self holds herself accountable. Elliott thus mistakes widespread, debased forms of authenticity in American culture for the morally robust ideal. As a result, he paints too bleak a portrait of enhancement in American culture. He may be right that, in a culture where it is hard to build an enduring sense of self, some—perhaps many—enhancement options are pursued for psychologically defensive reasons. [A choice or an action is psychologically defensive when it (consciously or unconsciously) protects the individual from difficult feelings or thoughts that—if reflected upon—might lead to a deeper and richer identity.] Yet in other cases, enhancement can be sought for the sake of a deeper and more genuine authenticity.

The appeal of enhancement technologies, as Elliott claims, does often reflect American anxieties about identity and the presentation of self in everyday life. Nevertheless, our culture also has deep and diverse—if often conflicting—religious resources to help us formulate a stance toward enhancement. And it also contains a more robust ideal of authenticity, in which self-fulfillment culminates in self-transcendence. Of course, neither religion nor authenticity guarantees justified and reasonable policies, practices, or regulations. But each is a valuable moral resource overlooked or underestimated by Elliott.

When de Tocqueville visited America in the 1830s, he was surprised by Americans' powerful commitment to religion—a commitment that shows no sign of abating in the twenty-first century. Elliott, on the other hand, follows the assumption of secularization built into almost all modernization theories—an assumption that contemporary sociologists of religion now acknowledge is quite misleading. America's diverse religious communities are an important and relatively untapped resource for addressing the moral and spiritual problems posed by medical enhancement.

In the setting of the Academic Coalition for Jewish Bioethics, we are pursuing discourse on these issues within our own tradition. In the larger American context, this should join into similar debates, perspectives, and interdenominational dialogue from Christian, Islamic, Hindu, and Buddhist theologians, preachers, and pastors. Each of these traditions contains frameworks of belief, hierarchies of value, and standards for living a good life, against which enhancement technologies can be evaluated.

4—"ANTI-AGING" MEDICINE AND THE CHALLENGES OF HUMAN ENHANCEMENT

The recent maturation of biogerontology is a striking example of the radical enhancement technologies that are emerging from biomedical research. The economic, social, and ethical dynamics of anti-aging interventions are outlined in an excellent article by Juengst, Binstock, Mehlman, Post, and Whitehouse, published in the summer 2003 issue of *The Hastings Center Report*. As they note, biogerontology is already struggling "with the prospect that its findings might be applied to produce unprecedented longevity" (21)—one of the most dramatic and far-reaching aspects of biomedicine's potential to re-engineer the human.

Biogerontologists are divided between those who labor to understand the basic mechanisms of aging in order to identify and remove pathologies and those who labor in search of anti-aging interventions. Scientists who embrace the goal of controlling aging suggest that we may be only a decade away from "engineered negligible senescence" or "aging reversal." Among serious biogerontologists, the majority fall into the first camp. But both camps of scientists work together to repudiate the growing clan of entrepreneurs and morally questionable clinicians who make false claims that effective "anti-aging" medicine has already arrived.

In terms of the treatment/enhancement debate, we might say that those who study the basic mechanisms of aging are seeking "treatments," while those who look for anti-aging interventions are seeking "enhancements." But as we noted above, the distinction is in certain ways socially constructed and artificial. At one level, this division reflects the struggle for funding and political boundary setting, with each camp claiming the authority of science (Binstock, 2003). In addition, recent experience suggests that public funding favors research that fights disease over research aimed at health promotion. In the private sector, however, investment seems to favor new biomedical enhancements over cures for disease. At a deeper level lie questions of the proper goals of research and the limits of efforts to unravel and control the basic biological processes of aging.

Most gerontologists agree that aging is not a disease but consists of biological processes that create the major risk factors for pathologies of old age.

As Leonard Hayflick puts it: "Aging is not a disease, [hence] seeking a cure for it is tantamount to seeking a cure for embryogenesis or child or adult development." But other researchers and their popularizers have no hesitation in seeking medical benefits and/or the potential "cure" of aging.

Complicating matters, the regulation of anti-aging products is affected by the persistent ambiguity over whether aging is a disease or a natural biological process. "If anti-aging products were limited to specific pathologies, they would count as drugs or medical devices, and proof of safety and efficacy would be required before they could be marketed. But if aging is not considered pathological, anti-aging products will be assessed like cosmetic medicines such as nonprescription contact lenses, breast implants, liposuction or Botox" (Juengst et. al., 2003). This pragmatic argument against retaining the therapy/enhancement distinction (that removing treatments from the category of "therapy" means exemption from FDA safety and efficacy regulations) can also be applied to other kinds of treatment. What is distinctive about anti-aging questions is the debate over whether the universal process of senescence can be categorized as pathological.

Although no currently marketed intervention has been proved to slow, stop, or arrest human aging and some are actually quite dangerous, the use of anti-aging products grew dramatically in the 1990s, and "anti-aging" medicine earned some of the trappings of legitimacy. There are now 11,000 members of the "American Academy of Anti-Aging Medicine," which campaigns aggressively against mainstream gerontology and geriatrics. In short, there is a large public that shares the dream of anti-aging and will quickly consume enhancement products and interventions.

Juengst and his colleagues draw three basic lessons from their research on anti-aging:

(1) They warn against the dangers of premature commercial exploitation of scientific findings.
(2) They encourage us to face the challenge of philosophical pluralism (i.e., the methods of herbal medicine, acupuncture, meditation, etc.) in medicine.
(3) They emphasize the necessity of open deliberations that anticipate rather than react to social repercussions of new medical technologies.

5—HOW SHOULD JEWS THINK ABOUT ANTI-AGING ENHANCEMENTS?

Aging American Jews—who make up more than 50 percent of congregational membership—are too often shaped by the American dream of unlimited accumulation of health and wealth. In subtle but powerful ways, retirement from work implicitly functions as a model of retirement from the covenant. Carl Elliott asks the key question: What vision of the good life do enhancement technologies serve? It must be said that contemporary Judaism lacks a persuasive vision of the good life for our elder years. Leisure clubs and life-cycle events become the default positions for engagement. The opportunities and responsibilities of age pale beside the dreams of eternal youth.

Fashioning authentically Jewish visions, images, and practices of later life is a serious priority today, though it is sadly neglected in both rabbinical education and congregational life. (Three notable American exceptions are Rabbi Address's "Sacred Aging" initiative at the Union for Reform Judaism; Rabbi Dayle Friedman's Center for Aging and Judaism at the Reconstructionist Rabbinical College in Philadelphia; and Rabbi Peter Knobel and Dr. Martha Holstein's project on Judaism, Aging, and Ethics at Temple Beth Emet in Chicago.) In the absence of any consensus or well-developed thought about these issues, we offer some thoughts to stimulate future discussion.

We would certainly not rule out all anti-aging interventions on the basis of any fixed idea of the human or of the natural lifespan. Where, then, does one turn for normative guidance? A fully developed answer will, of course, require a great deal of work in biblical and rabbinic sources. For an initial orientation, we look to Abraham Joshua Heschel's *Who is Man?* (1965), a slim volume based on the Raymond Fred West Memorial Lectures on Immortality, Human Conduct, and Human Destiny given at Stanford University in 1963. Barely twenty years after the Holocaust and World War II, Heschel described "the eclipse of humanity": "Today . . . the humanity of man . . . is no longer self-evident, and the issue we face is: How can a human being achieve certainty of his humanity?" (Heschel, 25).

Heschel's volume reminds us that "human being" is not identical to being human. Being human, as Heschel understood it, is a goal and a process that can be more or less successful. But becoming more human—a process that can only unfold with the passage of time—is especially difficult in a technological culture that instrumentalizes the self. The self becomes an object I can manipulate rather than an ever-changing inner experience in need of meaning and purpose. Hence, in the spirit of Rosenzweig's critique of metaphysics (Rosenzweig, [1921] 1999), Heschel does not ask the purely philosophical question, "What is man?" but rather the existential question, "Who is Man?"

We cannot simply "know" the answer to this question—we must live it. The Jewish tradition grapples with this question by affirming, in Heschel's words, "that to be is to be commanded." Jews belong to a covenant from which they cannot retire. But of course, that is precisely what many American Jews do—whether they belong to a congregation or not. They do not know how to live covenantally in later life.

For some older Jews, the synagogue provides an important source of identity and social support. Congregational leadership often includes influential elders with a lifelong commitment to the covenant. But many older Jews are detached from the tradition and do not feel its moral demands, sources of identity, comfort, and hope. And the broader culture squeezes older people into the limited roles of patient, consumer, and/or pensioner—effectively undermining a fuller notion of moral agency and responsibility—that is, of what it means to be an "elder" (Shalomi-Schachter and Miller, 1997).

In Jewish terms, we should begin to ask: What am I "called" to do or be? What does God require of me? We know that children are commanded to honor their parents. And there is a considerable literature on the responsibilities, difficulties, and limits of adult children caring for their parents (Greenberg, 1994; Address and Person, 2003; and see the discussions by Friedman and by Zimmerman in this volume). But what do older parents owe their adult children? What are the virtues and responsibilities of those who are frail and dependent? What kinds of blessing can those who are dying give to their loved ones? Aside from writing an ethical will, how should older Jews prepare themselves to participate actively in their own dying? Living as an exemplary Jewish elder might mean returning to (or finding for the first time) Torah study, finding meaningful forms of worship, visiting the sick, teaching, caring for grandchildren, etc.

These issues are now especially urgent, as the baby boom generation is becoming the "young old" and will soon become the "old old." Without some compelling prophetic or *halakhic* reconstruction of old age, American Jews will remain prey to the blandishments of merchants of immortality (Hall, 2003).

Abraham Joshua Heschel was also concerned about "authentic human existence" (Heschel, 1966, 75). He warned of the trivialized existence that emerges when ritual and prayer give way to a life of hobbies and recreation. In "To Grow in Wisdom," his classic address to the 1961 White House Conference on Aging, Heschel (1966) lamented the arid spiritual landscape of old age in a society preoccupied with possessing things and ill-equipped to sanctify time. He saw the Sabbath as a model for the celebration and sanctification of existence in old age—a model which I believe neglects society's need for participation and engagement in later life.

Without a radical culture change in our ideas about aging (as a mystery to be lived rather than a problem to be fixed), it is possible that life extension will actually contribute to a moral and spiritual rather than a biological dehumanization. If longer, enhanced lives emerge under the reigning regime of individualism, control, and the instrumentalized self, healthy older people may not become "more human," more spiritually and morally developed. In this case, one can foresee a gerontocracy of the rich and a tragedy of the commons (that is, a tragic loss for a society or community as a whole).

On the other hand, the Jewish concept of a *shomer* or caretaker (from the root sh.m.r.) offers an appealing ideal of dedicating ourselves to stewardship of our bodies in the service of God. In traditional Judaism, we do not own our bodies—they belong to God. Hence the duty to care for our bodies can be understood as a duty to the Owner of the body. As Moshe Chayim Luzzatto put it: "This too is a *mitzvah* upon us, to safeguard [*lishmor*] our bodies . . . in a fashion appropriate to enabling us to serve through it our Creator" (cited in Freedman, 1999, 176). If we keep the concept of *shomer* in our minds as we evaluate the enhancements of "anti-aging" medicine, it may carry us some way in deciding among various possibilities, as we continue to work toward an authentic Jewish vision of later life. From this perspective, "anti-aging" medicine may be used insofar as it enhances one's service to God—not insofar as it helps individual quest for unlimited health and wealth.

REFERENCES

Address, Richard and Hara E. Person, eds. (2003), *That You May Live Long: Caring for Our Aging Parents, Caring for Ourselves*, New York: Union of American Hebrew Congregations Press.

Best, Steven and Douglas Kellner, (1991), *Postmodern Theory: Critical Interrogations*, New York: Guilford Press.

Binstock, Robert (2003), "The War on 'Anti-Aging Medicine,'" *The Gerontologist*, 43, 1: 4–14.

Caplan, Arthur (2002), "No Brainer: Can We Cope with the Ethical Ramifications of New Knowledge of the Human Brain?" in *Neuroethics: Mapping the Field*, Conference Proceedings, San Francisco, CA, May 13–14, 96–106.

Daniels, Norman (2000), "Normal Functioning and the Treatment-Enhancement Distinction," *Cambridge Quarterly of Healthcare Ethics* 9, 309–22.

Elliott, Carl (2003), *Better Than Well: American Medicine Meets the American Dream*, New York: W.W. Norton.

Fukuyama, Francis (2002), *Our Posthuman Future: Consequences of the Biotechnology Revolution*, New York: Farrar, Straus and Giroux.

Freedman, Benjamin (1999), *Duty and Healing: Foundations of a Jewish Bioethic*, New York: Routledge.

Geertz, Clifford (1980), "Blurred Genres: The Reconfiguration of Social Thought," *American Scholar* 49, 165–79.

Giddens, Anthony (1991), *Modernity and Self-Identity*, Stanford: Stanford University Press.

Greenberg, Vivian (1994), *Children of a Certain Age*, New York: Lexington Books.

Hall, Stephen (2003), *Merchants of Immortality*, New York: Houghton Mifflin.

Harvey, David (1989), *The Condition of Postmodernity*, Cambridge, MA: Blackwell.

Heschel, Abraham Joshua (1965), *Who Is Man?* Stanford: Stanford University Press.

——— (1966), "To Grow in Wisdom," in *The Insecurity of Freedom*, New York: Schocken Books.

Jonsen, Albert R. (1998), *The Birth of Bioethics*, New York: Oxford University Press.

Juengst, Eric (1997), "Can Enhancement be Distinguished from Prevention in Genetic Medicine?" *Journal of Medicine and Philosophy* 22, 125–42.

Juengst, Eric, et al. (2003), "Biogerontology, 'Anti-aging Medicine,' and the Challenges of Human Enhancement," *Hastings Center Report*, 33 (4) (Jul./Aug.), 21 ff.

Kass, Leon (2002), *Life, Liberty, and the Defense of Dignity: The Challenge for Bioethics*, San Francisco: Encounter Books.

Lasch, Christopher (1978), *The Culture of Narcissism*, New York: Norton.

McGee, Glenn (2000), "Ethical Issues in Enhancement: An Introduction," *Cambridge Quarterly of Healthcare Ethics* 9, 299–303.

McKenny, Gerald and Estuardo Aguilar-Cordova (1999), "Gene Transfer for Therapy or Enhancement," *Human Gene Therapy* 10 (June 10), 1429–30.

Miller Franklin, Howard Brody, and Kevin Chung (2000), "Cosmetic Surgery and the Internal Morality of Medicine," *Cambridge Quarterly of Healthcare Ethics* 9, 353–64.

Murray, Thomas (2002), "Reflections on the Ethics of Genetic Enhancement," *Genetics in Medicine* 4 (6) (Nov./Dec.), 27S–32S.

Newell, Christopher (1999), "The Social Nature of Disability, Disease and Genetics: A Response to Gilliam, Persson, Holtug, Draper and Chadwick," *Journal of Medical Ethics* 25, 172–75.

Parens, Eric (1998), "Is Better Always Good? The Enhancement Project," *Hastings Center Report,* Jan.–Feb., S1–S15.

——— (2002), "How Far Will the Term Enhancement Get Us as We Grapple with New Ways to Shape Our Selves?" in *Neuroethics: Mapping the Field*, Conference Proceedings, San Francisco, CA.

Rabino, Isaac (2003), "Gene Therapy: Ethical Issues," *Theoretical Medicine* 24, 31–58.

Resnik, David and Pamela Langer (2001), "Human Germline Gene Therapy Reconsidered," *Human Gene Therapy* 12 (July 20), 1449–58.

Rieff, Philip (1966), *The Triumph of the Therapeutic*, New York: Harper and Row.

Riesman, David (1950), *The Lonely Crowd*, New Haven: Yale University Press.

Rosenau, Pauline Marie (1992), *Post-Modernism and the Social Sciences*, Princeton: Princeton University Press.

Rosenzweig, Franz (1999), *Understanding the Sick and the Healthy: A View of World, Man and God* (trans. Nahum Glatzer), Cambridge: Harvard University Press.

Rothman, David J. (1991), *Strangers at the Bedside*, New York: Basic Books.

Scully, Jackie Leach and Christoph Rehamann-Sutter, (2001), "When Norms Normalize: The Case of Genetic 'Enhancement,'" *Human Gene Therapy* 12 (Jan.1), 87–95.

Shalomi-Schachter, Zalman and Ron Miller (1997), *From Age-ing to Sage-ing*, New York: Warner Books.

Taylor, Charles (1992), *Ethics of Authenticity*, Cambridge, MA: Harvard University Press.

——— (1989), *Sources of the Self: The Making of Modern Identity*, Cambridge, MA: Harvard University Press.

Tocqueville, Alexis de (1840, 1945), *Democracy in America*, ed. Phillips Bradley, 2 vols., New York: Vintage Books.

Witrogen, McLeon Beth (1999). *Caregiving: The Spiritual Journey of Love, Loss and Renewal*. New York: John Wiley & Sons.

Wolpe, Paul Root (2002), "Neurotechnology, Cyborgs, and the Sense of Self," in *Neuroethics: Mapping the Field*, Conference Proceedings, San Francisco, CA, May 13–14, 159–67.

9

The Theological and Halakhic Legitimacy of Medical Therapy and Enhancement

Mordechai Halperin

My discussion here will follow the classical model of halakhic responsa, with two important differences. First, this is not truly a responsum—it was not written in response to an actual, real-life question, involving a particular person in specific circumstances. Its character is thus that of a hypothetical case study. Second, I will begin (after presenting the "case") with some methodological reflections, spelling out what might be implicit in standard halakhic writing.

A HYPOTHETICAL CASE

Moshe is twenty-three years old. Brought up in a mixed ethnic neighborhood, Moshe has long been self-conscious about his looks—particularly his nose, which he considers too big. He was teased about his nose as a child. He has recently been accepted to a prestigious law school, but before embarking on his legal studies he feels strongly that he should have cosmetic surgery to fix this feature of his face. He has tried to save money for school, but has little to spare, and may have to defer school for a year to pay for the surgery now.

Moshe's mother says he looks adorable as he is; she is concerned that his wish to shorten his nose expresses a deep desire to avoid looking "too

Jewish." Moshe argues that it is about feeling better about himself, and imagines that it would also increase his chances for a successful law career because he would "be more attractive to juries and to clients."

A rabbi Moshe consulted told him that since the surgery is by no means life-saving, and is not sought even to relieve pain or any real ailment, it does not constitute therapy. It is therefore not included in the "permission granted to the doctor to heal" and is halakhically impermissible—especially in light of the inherent risk, albeit not great. A second rabbi was also concerned about Moshe's possible assimilationist motives, but raised against this the consideration of marriage prospects—arguably the rhinoplasty will improve his chances of finding a mate. However, the rabbi also noted that a law degree from a prestigious school might be just as effective for that purpose.

What would be the appropriate halakhic guidance for Moshe? Would it make a difference if we replace him with "Miriam" in the same scenario? And what if, in addition to rhinoplasty, Moshe (or Miriam) wanted his dark hair dyed blonde, his skin bleached, and botox, to make him (or her) look more like the people they see in magazines and on television?

GENERAL POINTS—HALAKHIC METHODOLOGY

In this context, I wish to briefly address three points. First, what is the "default" halakhic stance—prohibition or permission? Second, who should render a halakhic decision? And third, how should halakhic analysis and decision proceed in the absence of clear precedents?

The *"Default"*: Prohibition or Permissibility?

The Mishnah (*Yadayim*, 4:3) emphasizes that only *prohibitive*, strict decisions require juridical substantiation while granting *permission*, or leniency, requires no supportive precedent. The absence of a prohibitive substantiation is to be equated with halakhic permissibility.[1] This implies that any technological innovation is permissible unless there is a halakhic reason for prohibiting it. If in the broad range of halakhic sources no reason is found for their prohibition, Jewish law permits the use of such technologies.

We may therefore conclude: The absence of a prohibitive substantiation is to be equated with halakhic permissibility.

Who Decides?

In order to be sure that there is no halakhic prohibition against a new procedure, an accepted halakhic authority must be consulted. Jewish law differentiates between the authority to abrogate a temporary prohibition and the authority to determine permanent permissibility. Faced with uncertainty or insufficient information, one is entitled to be strict with oneself; no special authority is needed for prohibition by the individual. On the other hand, in order to *establish* permissibility, there must be unequivocal information.[2] When there is no clear precedent in *halakhah* to decide the issue at hand, one must be thoroughly versed in all halakhic sources before definitely confirming that no halakhic reason for prohibition exists.

We may therefore conclude: An accepted halakhic authority must be consulted.

Issues without Clear Precedents

Step I: An attempt is made to find related precedents in halakhic literature. A possible result: no precedent.

Step II: halakhic study of conceptually connected rulings. A possible result: differences of opinion among the accepted authorities.

If there are no related precedents, halakhic study is made of conceptually connected rulings. In examining these, we attempt to infer the reasons upon which they are based. If these reasons are confirmed, or at least not contradicted, by other halakhic sources, they could be accepted for drawing conclusions regarding new issues under consideration.

Because of the vast range of halakhic material, there often arises a difference of opinions among the accepted authorities, though these differences are usually of short duration. Consensus is finally achieved and an unequivocal decision is reached.

Step III: using special halakhic rulings for controversial issues. There are well-known halakhic rules for deciding controversial issues. If, for example, there is a doubt in a matter prohibited by the Torah (*de-orayta*), the ruling is prohibitive; if the doubt is related to a rabbinical ruling (*de-rabbanan*), the decision is often permissive (see BT *Beitza* 3b).

HUMAN INTERVENTION IN THE AFFAIRS OF G-D

Medicine poses a fundamental question. In the Torah it appears that health is the divine reward for proper conduct. Suffering and disease are the punishment for sin and transgression:

> "If thou wilt diligently hearken to the voice of the Lord thy G-d, and wilt do that which is right in His sight, and wilt give ear to His commandments, and keep all His statutes, I will put none of these diseases upon thee, which I have brought upon Egypt: for I am the Lord that heals thee." (Exodus 15:26)

> "But if you will not hearken to Me, and will not do all these commands. . . . I will even appoint over you terror, consumption and fever, that shall consume the eyes, and cause sorrow of heart." (Leviticus 26:14–16)
>
> "And also every sickness, and every plague, which is not written in the book of this Torah, them will the Lord bring upon thee, until thou art destroyed." (Deuteronomy 28:61)

These verses seem to imply that medical treatment constitutes a gross interference in the divine scheme of reward and punishment. Even today, members of certain religions refuse all medical treatment so as not to interfere with "the will of God."

Halakhah, however, approves of medical treatment and sometimes considers it mandatory. The basis of the halakhic imperative to heal derives from the verse "Cause him to be thoroughly healed" (Exodus 21:19); our sages taught "Hence do we have permission to heal" (BT *Baba Kama* 85a, *Berakhot* 60a). From this it is derived that it is *incumbent* upon us to heal and save life, and withholding treatment is equivalent to shedding blood (SA, YD 336:1).

This special "permission to heal" is necessary, according to Rashi, since it might make sense to view illness as divine punishment, a process into which humans may not intrude. Nahmanides, in his book *Torat Ha-Adam*, likewise presents this theological explanation. In addition, he offers an alternative explanation, in terms of legal liability. Anyone who practices medicine will undoubtedly make some mistakes, despite best efforts. A medical mistake might result in a patient's death. Hence engaging in the practice of medicine entails a seemingly unbearable situation, wherein the practitioner risks killing with some degree of negligence. Therefore, absent the Torah's explicit permission, it might have seemed right to avoid this dangerous practice altogether. It was thus necessary for the Torah to permit this danger-ridden profession, so that overall healing and life-saving may be attained.

This unambiguous attitude of the *halakhah* regarding the obligation to heal calls for an explanation. If healing appears to represent an act of opposition to divine will, why should such intervention be permitted?

An early medieval *midrash* discusses this matter as follows:

> Rabbi Ishmael and Rabbi Akiva were walking in Jerusalem together with another man. A sick person met them and said: "Gentlemen, tell me how I may be healed." They responded: "Take such and such and you will be healed."
>
> After the sick person departed, the man who was accompanying the Rabbis asked: "Who caused his disease?" They answered: "The Holy One, blessed be He." He asked: "Why do you interfere in a matter which is not yours? The Lord did smite him; why then do you heal him?"
>
> The Rabbis asked him: "What is your occupation?" "I work the land. Here you can see my scythe," he answered. Then the Rabbis asked: "Who created the land upon which you work?" "The Holy One, blessed be He." "Then you are inter-

> fering in a matter which is not yours. The Lord did create the vineyard; why then do you eat His fruits?"
>
> The farmer responded: "Do you not see the scythe in my hand? If I did not plow and weed and put down fertilizer, nothing would grow in the land."
>
> "Fool," the Rabbis said, "a tree cannot grow if the land is not prepared. And if the tree grows, it will die unless fertilized and watered. Similarly the body of man must be tended by the physician with proper medication." (*Midrash Shmu'el 4; Midrash Temurah 2*)[3]

The idea expressed by this *midrash* is clear. The world was created with a system of natural laws. Humans are permitted to use the laws of nature to earn their livelihood and to maintain health. We may engage in farming for our livelihood, and it is appropriate to engage in medical therapy for our health. Human deeds of man do not detract from divine providence. Similarly, it is no offense to divine providence to give alms to the poor, for the Lord has many ways of providing for His creatures.[4]

Rabbi Abraham Ibn Ezra (twelfth century, Spain) had a seemingly maverick opinion on these matters. He distinguished between external injury perpetrated by humans, which one is permitted to treat, and internal disease caused by G-d, which one may not treat.[5] Although this opinion is not accepted by most authorities, it is important to understand Ibn Ezra's distinction between the different kinds of injury. Did he have a philosophical objection to human interference with internal disease caused by G-d? Or was his opinion based on experience that led him to the conclusion that internal injury is best left untreated so as not to endanger the patient with improper therapy, which was quite common in his day? Rabbi Elijah The Gaon of Vilna (eighteenth century), who was familiar with the standards of medical practice two hundred years ago, accepted the second explanation (Weinberger, M., pp. 11–34, n9).

PLASTIC SURGERY AND THE PERMISSION TO HEAL

On the basis of the above general considerations regarding medical practice, we can now begin to address the specifics of our hypothetical case. First let us consider the following argument against permitting the proposed surgery: The Torah granted "permission to heal" only for the sake of saving life or in order to avert significant impairment to a person's health. Hence, the permission does not extend to plastic surgery, which aims at neither of these goals.

In light of the above discussion, however, this argument appears ill-founded. Rabbi Akiva's response in the midrashic tale indicates that the permission to engage in medical practice is in principle analogous to the general permission to intervene in nature and to apply our knowledge of its laws for the sake of human well-being. Hence, once the permission has been granted, it is not restricted to medical treatment of disease; it extends to medical treatment to facilitate making a living, for example, just like the farmer harnesses nature to make a living.

An argument that would restrict the permission to intervene in nature—in the medical context, to purposes of life-saving (or the prevention or alleviation of serious impairment)—is not consistent with the theological approach delineated above, and in fact is not endorsed by the majority of halakhic decisors.

A second possible ground for prohibiting plastic surgery would be the prohibition of inflicting injury. A person is prohibited from injuring not only others but also him- or herself, as stated in the Mishnah (*Bava Kama* 8:6). Halakhic literature contains numerous discussions of this question. For present purposes, it seems sufficient to cite a responsum by Rabbi Moshe Feinstein (late twentieth century, New York):

That discussion presents weighty halakhic reasons permitting plastic surgery for real needs, such as a young woman who needs such surgery in order to increase her chances of marriage. I reproduce here some excerpts from this responsum:

> [Dated] 20th of Adar 5724 (=1964)
>
> I was asked about a young woman who wishes to make herself beautiful, so that [many] will seek her hand in marriage, by means of the doctors' recent invention—surgery that constitutes injury to her body. Is she permitted, in light of the prohibition of self-injury? Note that the Tosafists (*Bava Kama* 91b, s.v. "Ela") indicate that self-injury is prohibited even for a [legitimate] purpose. . . .
>
> Maimonides, however, defined the prohibition of injuring another Israelite as requiring an "aggressive manner"[6] (Laws of Injury 5:1). . . . Hence in the case before us, where the injury is in order to make her beautiful, it is not in an aggressive manner and the prohibition does not apply. If [this qualification] applies with regard to injuring another, the same holds regarding self-injury. . . .
>
> She should therefore be permitted to make herself beautiful, even though it is achieved through injury, since it is not in an aggressive manner but on the contrary, for her own good. . . .
>
> Indeed, if the injured person consents, it is permitted even without relying on [the qualification stated by] Maimonides. . . . If it is done for the person's own good and with his [or her] consent, [this can be deduced] from the verse, "Love your neighbor as yourself" (Lev. 19:18) . . . especially with regard to a young woman, for whom becoming beautiful is more of a need and a benefit than for a man. (*Responsa Iggerot Moshe, Hoshen Mishpat* 2:66)

Thus according to Rabbi Feinstein, the prohibition of "inflicting injury" does not apply if the following two conditions are both met:

(1) The injury is induced with the injured person's consent.
(2) The injury is objectively for that person's own good.

From this responsum we can deduce that plastic surgery can be likewise permitted where a person might find himself unable to make a living, if an objective assessment determines that plastic surgery can improve his chances.

It is necessary to address one other argument against permitting plastic surgery: the prohibition upon an individual to place him- or herself in danger. Indeed, every surgery involves risk—albeit small—of death resulting from anesthesia or complications. Some assessments put the risk of mortality from anesthesia, for a healthy person, at between 1:10,000 to 1:20,000, roughly the same as the annual risk of death in a traffic accident.

Despite the prevalence of this consideration in medical-halakhic discussions, it carries little weight insofar as the risks are minuscule and the goals at stake are important ones, such as enhancing one's ability to make a living or chances of marriage.

Rabbi Yehezkel Landa (Prague, late eighteenth century) relied on the talmudic discussion in tractate *Bava Metsia* 112a that it is permissible to undertake risks for the sake of making a living, provided that the risk is relatively small. As is well known, a similar point was made by Rabbi Feinstein and in other halakhic works.[7]

Hence if the plastic surgery can in fact facilitate finding employment, its low-level risk does not provide halakhic grounds for a prohibition—provided, of course, that the individual is properly informed and consents to the actual risks of bad results and of complications.

SPECIAL ASPECTS OF THIS CASE AS DESCRIBED

We must attend to the possibility that the proposed surgery is—perhaps unconsciously—a first step toward assimilation. If so, endorsing this act might amount to a contribution to the future process of assimilation, and thus constitutes a transgression of the injunction, "Do not place a stumbling block before the blind." It is also possible, however, that a laconic prohibitive answer—without any attempt to forge an emotional connection to this Jewish student—may equally cause alienation and contribute to a process of assimilation. Thus one who offers that kind of response would also be transgressing "Do not place a stumbling block before the blind."

It is appropriate to cite here the Sages' teaching:

> Our Rabbis taught: While the left hand should push away, the right hand should draw near not like Elisha, who pushed Gehazi away with both hands, and not like Joshua ben Perahiah who pushed away one of his students[8] with both hands. (BT *Sotah* 47a)

Implementing this rabbinic instruction, "While the left hand should push away, the right hand should draw near," requires personal sensitivity and much wisdom and commitment. Not everyone can properly follow it. But whoever is responsible for the community in which such a student resides must make the utmost effort to preserve the "Jewish spark" in the heart of this individual, for it is that spark that might prevent assimilation.

In this regard it makes no difference whether we are discussing plastic surgery or dyeing one's hair, a young man or a young woman. Each case must be examined in light of the individual's particular circumstances. In general terms, we should follow the halakhic principle formulated in the Jerusalem Talmud: "[J]ust as it is forbidden to pronounce 'permitted' that which is prohibited, so too is it forbidden to pronounce 'prohibited' that which is permitted." (This principle is cited by Rabbi Joseph Caro in *Bet Yosef*, YD 115.) The danger of assimilation requires not only extra care in rendering a halakhic ruling, but also personal commitment and an ongoing effort. This commitment must be undertaken not only by the rabbi, but also by all adult community members. Together, they should work to strengthen the Jewish spark and the emotional ties of young individuals to their ancestral tradition and to their community.

REFERENCES

Weinberger, M. (1985). "Call for Medical Help According to *Halakhah,*" in M. Halperin, (ed.) *Emek halakhah-Assia*, Jerusalem: Schlesinger Institute.

NOTES

1. *Tiferet Israel, Yadayim*, 4:3; Rabbi E. Wassermann, *Kovetz He'arot Yevamot* 87b, section. 67, (550), based on Nahmanides and Adret.

2. See Rashi, *Beitsa* 2b s.v. "De-Hetera."

3. Cited in *Sefer ha-Pardes;* cf. C. Kahn's introduction to *Sefer Assia*, Vol. 2, ed. A. Steinberg (Jerusalem: Reuven Mass, 1981), p. 5.

4. Maimonides, Commentary on the *Mishnah, Pesahim*, at the end of chapter 4.

5. Ibn Ezra's Commentary on Exodus 21:19.

6. Rabbi Feinstein notes that some versions read "contemptuous" instead of "aggressive"; in his ensuing discussion, he points out that the case in hand involves neither of these abusive characteristics.

7. See *Responsa Noda Bihuda 2, Yoreh Deah* 10, and *Responsa Iggerot Moshe, Hoshen Mishpat* 1:104.

8. The words "one of his students" were substituted, on account of censorship, for the original "Jesus of Nazareth."

10

Therapy and Enhancement: Jewish Values on the Power and Purpose of Medicine

Louis E. Newman

The question before us—how to distinguish between therapy and enhancement—is complex and confusing. We all recognize, and Jewish tradition wholeheartedly affirms, that medical technology can and frequently must be used to heal the sick and repair the physical and psychological ailments that all of us suffer from. Yet, medical science can also be used for enhancing our bodies even when we are not ill, as we know from the use of steroids by athletes and from the example of plastic surgery. With rapid advances in genetic medicine, we are now or soon will be able to enhance our physical abilities and very likely extend our longevity significantly beyond anything we could have done in the past. And this latter possibility is troubling because it appears to involve tampering with what is natural or normal. Yet, defining the boundary between what is natural and what is not, what is normal for the human species and what is "enhanced," turns out to be a vexing problem, indeed. (After all, vaccines are designed precisely to give our bodies enhanced abilities to fend off diseases that we would get otherwise, and few of us would regard vaccinations as illegitimate.)

I want to suggest that this issue, like so many others raised by medical technology, is at root a question of how we use our power (in this case the power of medicine) to enhance and even transform the capabilities of our bodies—to make ourselves stronger, or smarter, or to enhance the capacities of our senses or our minds. The first rule of ethics is that all power comes with responsibilities. The difficulty is in discerning the specific risks that come with a given kind of power, and so the responsibilities and limits that should attend the exercise of that power. And if the power in question is unprecedented, or vastly greater than similar powers we have enjoyed in the past, we rightly sense that we are "out of our depth" and lacking in clear guidance. I will not take the time here to review the many proposals that have been made for distinguishing between "therapy" and "enhancement." Suffice it to say, this is a very difficult distinction to draw clearly.

What I shall offer here can be characterized as a "values" approach—an approach that I regard as the best way to unravel these moral quandaries. Let me begin, then, with a word about values. We do well to think about values in terms of goals or end-states. If I say that I value justice, I mean that I seek to create a world in which judges are unbiased and impartial, and in which everyone has an equal opportunity to share in the goods and opportunities that society offers. If I say that I value compassion, I mean that I seek to create a world in which people help those who are underprivileged or marginalized, and in which people strive to forgive those who have harmed them. Values, in short, can be defined in terms of the goals we seek, the sort of world we wish to create through our moral deeds. To take a "values-based" approach to a moral problem, then, is to consider the end-state that we wish to achieve, the goal that we "value," and then to act so as to maximize that outcome.

Problems arise, however, when we face a situation in which we value two different goals and life conspires to make it impossible for us to achieve both. Justice may demand that a convicted murderer receive a harsh sentence, while compassion may demand that we seek to empathize with and forgive him. In such cases, to work toward one goal is necessarily to impede our progress toward another. In short, we face a conflict in values and we often lack a clear sense of how to determine which value, which goal, should have priority. We cannot serve two masters, at least not equally. Such is the case with the therapy/enhancement debate, I think. On the one hand, we value the power of medicine to maximize our healthy functioning. On the other hand, we value a world in which we are roughly equally endowed, or at least in which physical advantages are distributed among us randomly, not available for purchase by those with the monetary resources to avail themselves of the latest "enhancements." We cannot have it both ways, and we are uncertain how to mediate between the conflicting values that make claims upon us.

Within Judaism's vast corpus of moral teachings, we find values most clearly articulated in the non-halakhic, non-legal sources—*midrash*, liturgy, aphorisms—these are the primary sources of Jewish values, because they provide us with stories or visions of how the world ought to be, of what ends we ought to seek and how to pursue them. (In saying this, I do not at all mean to denigrate the importance of *halakha* to Jewish ethics. I only wish to note that legalists will approach a moral problem with an eye to finding the rule or precedent that is most applicable, rather than with an eye to finding the value that should be maximized, or the best way to mediate among conflicting values.) In the remainder of this essay, I want to introduce a few midrashic and liturgical texts that I think capture these conflicting values. For, as we will soon see, Jewish texts concerning medicine, healing, and the human body reflect the very tension that I think we all feel when we face the vexing questions of distinguishing between legitimate therapy and illegitimate enhancement. And yet, I also hope to show that there is at least one overriding Jewish value that helps us to mediate this conflict and points toward a resolution of this problem.

The first text I want to introduce comes from *Sifre Devarim*, a rabbinic *midrash* on the book of Deuteronomy:

> Rabbi Simeon ben Yohai, quoting 'The Craftsman, whose work is perfect' (Deut. 32:4) said: The Craftsman who wrought the world and man, His work is perfect. In the way of the world, when a king of flesh and blood builds a palace, mortals who enter it say: Had the columns been taller, how much more beautiful the place would have been! Had the walls been higher, how much more beautiful it would have been! Had the ceiling been loftier, how much more beautiful it would have been! But does anyone come and say: If I had three eyes, three arms, three legs, how much better off I would be! If I walked on my head, or if my face were turned backward, how much better off I would be! I wonder. To assure that no one would say such a thing, the King of kings of kings, the Holy One and His court had themselves, in a manner of speaking, polled concerning the placing of every part of your body and set you up in a way that is right for you. (*Sifre Deuteronomy* 307)[1]

One cannot read such a text today, in light of the promises of genetic technology, without experiencing a touch of irony. No one, perhaps, would set out to design a person with three eyes or a face turned backwards. But a great many people do indeed wish to redesign people to be "better" than they are by nature—stronger, less susceptible to disease, more of one thing, less of another. Our text suggests that such efforts are foolhardy, for we are already the products of a "master designer"; all of our essential needs can be met without altering the form in which we have been created. And in the contrast drawn to the way we respond to human creations, the text further suggests that we ought not to regard our bodies as akin to a human artifact. Artifacts we design to our liking, and we can readily re-design them to better suit our needs. But our bodies are not of our own making; hence, we ought not to indulge in idle speculation about how we might improve on them.

It should be noted that this text does not seem to address the problems of those born with physical deformities, but only the normal physical traits of our species. We know from other sources that the rabbis did not object to altering or curing physical ailments in those who are abnormal. So, we can infer that they mean to condemn only the effort to improve upon the characteristics of humankind as a whole. Any such effort would imply that we are better "craftsmen" than God, which is precisely what this *midrash* wants to refute.

Viewing God as the master craftsman of our bodies could inform a very skeptical attitude toward genetic enhancements, but before pursuing this line of thinking, we do well to consider another midrashic text that points in a rather different direction.

> It is told of Rabbi Ishmael and Rabbi Akiva that, while they were walking through the streets of Jerusalem accompanied by a certain man, a sick person confronted them and said, 'Masters, tell me, how shall I be healed?' They replied, 'Take such-and-such, and you will be healed.' The man accompanying the sages asked them, 'Who smote him with sickness?' They replied, 'The Holy One.' The man: 'And you bring yourselves into a matter that does not concern you? God smote, and you would heal?' The sages: 'What is your work?' The man: 'I am a tiller of the soil. You see the sickle in my hand.' The sages: 'Who created the vineyard?' The man: 'The Holy One.' The sages: 'Then why do you bring yourself into a matter that does not concern you? God created it, and you eat the fruit from it!' The man: 'Don't you see the sickle in my hand? If I did not go out and plow the vineyard, prune it, compost it, and weed it, it would have yielded nothing.' The sages: 'You are the biggest fool in the world! Have you not heard the verse, 'As for man, his days are as grass,' (Ps. 103:15)? A tree, if it is not composted, weeded, and [the area around it] plowed, will not grow; and even if it does grow, if not given water to drink, it will die—will not live. So, too, the human body is a tree, a healing potion is the compost, and a physician is the tiller of the soil.' (*Midrash Samuel* 4)[2]

Here the operative metaphors—the human body as a plant, the physician as a farmer—point toward a more active stance in relation to the body's needs. We must "cultivate" our bodies, intervening in the natural process to ensure that they develop properly. There is here a kind of interplay between what is naturally given (earth/plant, or the human body) and the human activity of developing that (through agriculture or medicine). The rabbis' interlocutor is made to look simple-minded, though on one level he simply represents a different theological perspective, albeit one that the rabbis have forcefully rejected. From that man's perspective, what God has created should be left in God's hands.[3] The rabbis challenge that view through a kind of *reductio ad absurdum*. To preclude people from interfering in God's world entirely would greatly impede normal human development, just as refraining from tilling the soil would impede the growth of a tree. So, too, when we intervene in the natural processes of the human body, we are only "pruning away" the elements that preclude it from developing in the way that God intended. One could also read this *midrash* more restrictively, as sanctioning such interventions in the natural order *only* when they are necessary for sustaining human life, as in the case of agriculture. That reading would be supported by the reference toward the end of the passage to the threat that without intervention the plant (or person) will die. Where human life is imperiled, intervention is required; we do not uniformly prohibit tampering with nature on the grounds that it is God's domain.

In light of these two contrasting sources, the moral challenge we are addressing today might be cast as that of finding some equilibrium between (1) accepting and honoring our bodies just as they are, as the work of a Master Craftsman, and (2) "cultivating" our bodies, nourishing and pruning them to maximize their potential for healthy development (or, at least, to preserve life). Both sources make powerful claims upon us. The former instructs us that we are, after all, God's creation and we dishonor our Creator if we presume to improve fundamentally upon God's work. The latter reminds us that human life requires us to intervene in the workings of nature, which include the processes of the human body; we are partners with God, "cultivating" and "tending" the raw material that God has given us.

Into this pair of conflicting values, I wish to introduce yet another text, this time a blessing which Jewish tradition requires that we recite upon seeing someone physically deformed:

> Blessed are you, Lord our God, who fashions diverse creatures (*m'shaneh ha-b'riyot*).[4]

The notion of reciting a blessing praising God for something that is "abnormal" may seem odd, even perverse. The value implicit in the practice of reciting such a blessing is worth examining more closely. All that occurs naturally is from God. The same Creator who is the source of rainbows and cloudless, sunny days is also the source of devastating storms and earthquakes. More to the point, God is behind both the perfectly healthy, wondrously functioning human body and the body that is naturally impaired, less than whole and healthy. The tradition of reciting this blessing, then, challenges us to set aside, at least temporarily, our preconceived notions about what is "normal" vs. "abnormal" and, more importantly, to abandon our instinctive inclination to value the former and devalue the latter.[5] If we were to internalize the message of this blessing, we would come to affirm that God is at work in all of creation, including those parts that seem to us "dysfunctional."

And yet, of course, reciting this blessing in no way mitigates our obligation to heal the sick, or to remove such impediments to healthy human functioning as can be removed. But the goal of healing is not that healthy, functioning bodies are "better" or "more blessed" than malfunctioning ones, for otherwise there would be no point to the blessing. The goal and meaning of healing is decidedly *not* about making our bodies "better" or "more perfect," more like some ideal represented by a Greek statue or (to use a more contemporary example) a Hollywood model. All bodies, no matter how impaired, are occasions for praising God as creator of all. We come, then, to the critical question: How are we to understand the responsibility to heal, the power of medicine and its limits, from a Jewish values perspective?

I think Maimonides best captured the essence of the classical Jewish view of health and healing when he wrote:

> He who regulates his life in accordance with the laws of medicine with the sole motive of maintaining a sound and vigorous physique and begetting children to do his work and labor for his benefit is not following the right course. A man should aim to maintain physical health and vigor in order that his soul may be upright, in a condition to know God. . . . Whoever throughout his life follows this course will be continually serving God . . . because his purpose in all that he does will be to satisfy his needs so as to have a sound body with which to serve God. Even when he sleeps and seeks repose to calm his mind and rest his body so as not to fall sick and be incapacitated from serving God, his sleep is service of the Almighty.[6]

Medicine is valuable because it is essential to the human flourishing that alone enables us to fulfill our divine mission. In order to do God's work in the world, we must first and foremost make every effort to preserve life against any and all forces that threaten it. Moreover, we must ameliorate any condition that diminishes our ability to observe God's law, and to do deeds of lovingkindness. In short, the human body and its health are valued, not as ends-in-themselves, but as the necessary means to the end of serving God and bringing more holiness into the world.[7] Dysfunction and pain—whether physical or psychological—make this work more difficult and, in extreme cases, impossible. Healing the sick, then, is just a way of ensuring that our ability to serve God is restored, as much as that is possible.

But how does this perspective on health and healing shape our response to the therapy vs. enhancement issue that we face? I want to suggest that, if medicine as a whole is meant to enable us to serve God, it follows that any particular medical therapy is valid only to the extent that it furthers this purpose. A medical therapy that made serving God more difficult would be self-defeating. It might serve our bodies, but only at the cost of ignoring the fact that our bodies are meant to serve God. The ultimate goal of Jewish moral life, the highest value in Judaism's moral system, is reverence for God. The world that we strive to create through our moral action is, in the classic phrase, "the kingdom of God on earth," a world in which all recognize and revere God's sovereignty. It is that value that helps us to mediate between the conflicting values we noted earlier. Our purpose on earth is to live reverently, serving God in all we do. Sometimes this requires that we "cultivate" the body that God has given us; at other times, it requires that we honor the natural order of things just as God has created it. Just as we do God's bidding both when we labor for six days and when we rest on the Sabbath, so too we must distinguish between the times when we serve God through medical interventions and those when we accomplish the same purpose through refraining from such manipulations of our natural state.

This criterion, broad as it is, helps us to distinguish between those genetic therapies that are morally required and those that are forbidden. In general, we are required to provide any therapy that will either (a) save life, or (b) restore those basic capacities required for the performance of righteous deeds. Those capacities include, of course, the abilities to walk, talk, hear, and see, as well as the mental capacities to think, feel, plan, and cooperate with others. A person lacking one or more of these capacities will face significant impediments to doing deeds of lovingkindness.[8] From the perspective articulated by Maimonides, medical intervention is morally required when it is necessary to enable people to do what is required of them to serve God through righteous deeds.

On the other end of the spectrum, certain sorts of therapies (which we usually think of as "enhancements") are plainly precluded by this criterion. Any therapy used to give someone a physical or mental capacity far beyond that which is required to perform deeds of lovingkindness is prohibited. This would include the oft-discussed efforts to give people extraordinary athletic or artistic or intellectual endowments. While it could certainly be argued that such therapies would exponentially increase one's abilities, including the ability to do God's work in the world, such powers are not required for that purpose. To put it starkly, having more great athletes or musicians in no way furthers the moral and religious goals for which we were created, which is to do deeds of righteousness and holiness.[9]

It is important to note that this criterion cannot be applied on the basis of the motive of the patient or even the likelihood of a therapy being used in a specific way. The patient with spinal cord injury who can (hypothetically) be cured with stem-cell therapy may be motivated by the possibility of someday running a marathon. Still, since that therapy removes a major impediment to the patient's mobility, and so makes it easier to perform good deeds, it is required, notwithstanding the patient's less than noble goals (from a religious perspective). Similarly, the person who is functionally blind, but whose sight may be restored through medical intervention, may use that sight to shoot a high-powered rifle at innocent civilians. It is not the physician's job to pass judgment on the moral motivations of the patient, or to assess the probability that the patient will use his or her newfound capacity for good or ill. From the religious standpoint adopted here, the critical issue is whether the therapy is necessary to give the patient those capacities needed to live a full moral life.

The theological perspective I have been advocating here points to two further issues that deserve attention. First, since only the living can serve God, any genetic therapy that saves life or prolongs it is necessary.[10] One of the major promises of genetic therapies is that of greatly extending human longevity, and such a development would be most welcome from this values perspective. The longer we live (assuming continued health and vitality), the greater our opportunities to perform deeds of lovingkindness. Second, if further research into the human genome were to identify a specific gene that encouraged violent behavior, and which seemed to have no socially positive effect, I suggest that, based on the values outlined here, it would be morally appropriate (and perhaps even required) to alter this gene (assuming, of course, that this could be done safely, that we knew the long-term consequences of this change, etc.). This would constitute a medical intervention in the name of creating people more prone to righteous behavior (or, at least, less prone to destructive behavior), and this would put the technology directly in the service of living a godly life.[11]

In closing, I want to note one important way in which this values orientation is similar to a more halakhic orientation to Jewish ethics. As a practical matter, neither traditional Jewish values nor traditional Jewish laws bind us unless we choose to be bound. It is one thing to discover the core values that animate Jewish discussions of health and healing; it is quite another to adopt those values as our own. Indeed, I suspect that many of us (and I include myself in this description) will feel some degree of discomfort with the idea that such specifically theological values should guide our decisions about the proper uses of our medical technology. But this is just as it should be. For most contemporary Jews (perhaps, for most contemporary religious people of all persuasions), we delve into the wisdom of our respective traditions not because we are committed to adopting wholesale everything that we find there. We do so, instead, precisely because it often challenges us to consider values quite different from our own. Confronted with traditional Jewish values, we may be stimulated to re-consider our own, to reflect on what goals and end-states we really do wish to pursue. In the case before us, we may come to realize that what is at stake is not merely how we distinguish therapy from enhancement, or even how we understand the goals of medicine in general, but rather how we view the goals of human life as a whole. In doing this, we begin to frame our moral options differently and perhaps to weigh the conflicting values we espouse differently, as well. And it is precisely by considering alternative sets of values that we enrich our sense of moral imagination, hone our ability to discern the relative merits of the various values we might choose among, and deepen our appreciation of the responsibilities that come with our awesome power to heal.

REFERENCES

Bialik, Hayim Nahman and Ravnitzky, Yehoshua Hana, eds. (1992), *The Book of Legends [Sefer Ha-aggadah]*, New York: Schocken.

Twersky, Isadore, ed. (1972), *A Maimonides Reader*, New York: Behrman House.

NOTES

1. Cited in Bialik and Ravnitzky (1992), 14. A variation of the same text appears in Genesis Rabbah 12:1. The *midrash* plays on the word "tzur" ("Rock") in Deut. 32:4, reading it instead as "tzayyar," ("Craftsman").

2. Cited in Bialik and Ravnitzky (1992), 594–5.

3. The rabbis skillfully point out the man's inconsistent application of this theological principle. But we need only consider the views of contemporary Christian Scientists with respect to medical interventions to appreciate that such a view can be held consistently.

4. BT *Berakhot* 58b (my translation).

5. The same theological point, I think, is expressed through the notion that "one must bless God over the evil [that occurs to us] just as one blesses God over the good." See Mishnah *Berakhot* 9:5.

6. Maimonides, *Mishneh Torah, Hilkhot De'ot* 3:3, as cited in Twersky (1972), 57.

7. It is standard in many presentations of Jewish healthcare ethics to suggest that the preservation of life is an absolute value in Judaism, not relative to any other purpose or goal. But classical sources affirming the value (in extreme cases, even the requirement) of martyrdom belie such simple formulations. In certain cases, one is required to sacrifice one's life for a higher good, that of serving God and fulfilling the most important of God's commandments. The same is true, I suggest, in the realm of health care.

8. Note that "disability" and "impediment" are always matters of degree. The clearest moral cases are those in which the disability involved makes it impossible or very nearly impossible to do that which God put us on earth to do. There will always be "borderline" cases in which medical intervention can incrementally improve a patient's functioning. In such cases, it will always be a judgment call as to whether the impediment suffered by the patient is significant enough to warrant removal.

9. Between the extremes of that which is morally required and that which is morally prohibited lies that all-important and ambiguous area of the permissible. In the case of genetic therapies, that territory might be defined as including incremental enhancements that are not necessary to restore basic physical or mental functions, but which also are not used to produce supernatural abilities. Here presumably other criteria will need to be invoked, including the needs and history of the individual patient, and the social ramifications of making such therapies available to some patients, but not others. It seems to me that Jewish ethicists, including those who adopt the approach I advocate here, could legitimately differ on these borderline cases.

10. This presumes that the threat to life is both immediate and reversible. Many Jewish authorities have held that when death is imminent and unavoidable, we are

permitted to let events run their course. This position is consistent with the view advocated here.

11. Some could object here that, if we were genetically "programmed" to be righteous, we would lose the merit that we accrue by overcoming our natural evil instinct in order to do good. On this view, part of the "value" of doing righteous deeds is that, for most of us, it *doesn't* come "naturally," we have to work at it. But I contend that Jewish authorities have generally been more concerned about consequences than about motivation, at least in the realm of ethics. The person who gives charity, even for the wrong reason, is to be preferred over the person who does not. By the same token, the tradition would value the person who lives righteously over the violent sociopath, irrespective of the circumstances that produced this behavior. Moreover, as Noam Zohar has suggested to me, one could argue that violent sociopaths are victims of inner compulsions they cannot control, so genetic therapy would really serve to give them the free will necessary to serve God. Again, all this assumes that there is some specific genetic basis for extremely violent behavior, that it could be altered, that this could be done safely, and that there were no adverse consequences for the physical or mental health of the individual involved. I am grateful to Dr. Alan Astrow for his thoughtful probing of this point.

Index

About the Contributors

Thomas R. Cole, Ph.D., is the Beth Toby Grossman Professor in Spirituality and Health and Director of The John P. McGovern, M.D., Center for Health, Humanities, and the Human Spirit at the University of Texas School of Medicine in Houston. He is the author of *The Journey of Life* and author or editor of many articles and several books on the history of aging, humanistic gerontology, and medical humanities.

Rabbi **William Cutter**, Ph.D., is Steinberg Professor of Human Relations and Professor of Education and Modern Hebrew Literature at the Los Angeles Campus of Hebrew Union College Jewish Institute of Religion, and also directs the College's Lee and Irving Kalsman Institute on Judaism and Health. He has written numerous articles on literary theory, modern Hebrew literature, and the conjunction of education, health, and literature. Dr. Cutter is President of the ACJB.

Rabbi **Elliot N. Dorff**, Ph.D., is Rector, Distinguished Professor of Philosophy, and Co-Chair of the Bioethics Department at the University of Judaism. He has served on three federal government commissions on, respectively, health care accessibility, responsible sexual conduct, and protecting human subjects in research, and he is now a member of the Ethics Advisory Committee for the

California Stem Cell Project. Vice Chair of the Conservative Movement's Committee on Jewish Law and Standards, he serves as President of both Jewish Family Service of Los Angeles and the Society of Jewish Ethics. He is author of over 150 articles and twelve books on Jewish thought, law, and ethics, including *Matters of Life and Death: A Jewish Approach to Modern Medical Ethics* (1998).

Rabbi **Dayle A. Friedman**, M.A.J.C.S./B.C.J.C. M.S.W., is the founding director of *Hiddur*: The Center for Aging and Judaism at the Reconstructionist Rabbinical College, which works to transform the culture of aging in the Jewish community. She also serves as a clinical supervisor and a member of the faculty at RRC. She is editor of *Jewish Pastoral Care: A Practical Handbook from Traditional and Contemporary Sources* 2nd ed. 2005. Rabbi Friedman formerly served as founding Director of Chaplaincy Services at Philadelphia Geriatric Center, where she created the first clinical training in aging for rabbinic and cantorial students, and initiated and co-chaired the Center's Medical Ethics Committee.

Rabbi **Mordechai Halperin**, M.D., serves as the Chief Officer of Medical Ethics in the Israeli Ministry of Health and directs the Falk Schlesinger Institute for Medical Halakhic Research at the Shaare Zedek Medical Center in Jerusalem. A board member of the Bioethics Advisory Committee of the Israeli Academy of Arts and Sciences, he was formerly the Director of the Jerusalem Medical Center for Impotence and Infertility. Dr. Halperin edits *Assia*, the Hebrew Quarterly Review of Medical Ethics and Jewish Law, as well as the English-language journal, *Jewish Medical Ethics* (JME).

Louis E. Newman, Ph.D., Musser Professor of Religious Studies at Carleton College, Northfield, Minnesota, also chairs the Religion Department and directs the program in Judaic Studies. He is the author of *Past Imperatives: Studies in the History and Theory of Jewish Ethics* (1998) and co-editor, with Rabbi Elliot Dorff, of *Contemporary Jewish Ethics and Morality* (1995) and *Contemporary Jewish Theology* (1998). Dr. Newman recently served as the first President of the Society of Jewish Ethics.

Robin Solomon, M.A., is a Ph.D. candidate in Medical Humanities at the University of Texas Medical Branch in Galveston. She holds a Master's degree in Bioethics from the Medical College of Wisconsin and continues her career in education where she began as a secondary science teacher.

Paul Root Wolpe, Ph.D., is a Senior Fellow of the Center for Bioethics and associate professor at the University of Pennsylvania, where he holds faculty appointments in the Departments of Psychiatry, Medical Ethics, and Sociology. He serves as the first Chief of Bioethics for the National Aeronautics and Space Administration (NASA), and is the first National Bioethics Advisor to the Planned Parenthood Federation of America. Dr. Wolpe is President of the American Society for Bioethics and Humanities, and is the Associate Editor of the *American Journal of Bioethics*. The author of numerous articles and book chapters in sociology, medicine, and bioethics, Dr. Wolpe does research that examines the role of ideology, religion, and culture in medical thought, especially in relation to emerging biotechnologies.

Deena Zimmerman, M.D., M.P.H., I.B.C.L.C., known to thousands of observant Jewish women throughout the world in her capacity as a *yoetzet halakhah* (consultant in Jewish law), is a graduate of the Keren Ariel Program of Nishmat, the Jerusalem Center for Advanced Jewish Study for Women. She is the author of *A Lifetime Companion to the Laws of Jewish Family Life* as well as a number of articles on medicine and *halakha*. She is the coordinator of the Nishmat's Women's On Line Information Center (<http://www.yoatzot.org> and <http://www.jewishwomenshealth.org>). Dr. Zimmerman practices as a pediatrician and lectures widely on women's health issues and *halakhah*, and on the promotion and support of breastfeeding.

Rabbi **Noam J. Zohar**, Ph.D., is a member of the philosophy department at Bar Ilan University and Director of its Graduate Program in Bioethics. He is Senior Research Fellow at the Shalom Hartman Institute in Jerusalem, and a member of the Ethics Committee of Rabin Medical Center, Petah Tikva. Dr. Zohar has published articles in the fields of Rabbinics and of moral and political philosophy. He is the author of *Alternatives in Jewish Bioethics* (1997) and (with Michael Walzer and others) *The Jewish Political Tradition* (Volume 1: Authority, 2000; Volume 2: Membership, 2003).

Laurie Zoloth, Ph.D., is the Director of Ethics at the Center for Genetic Medicine of the Feinberg School of Medicine at Northwestern University and Professor in The Program of Medical Humanities and Bioethics at the Northwestern University and in the Department of Religion at Northwestern. Former President of the American Society for Bioethics and Humanities, she serves on the Executive Committee of the International Society for Stem Cell Research and is a member of NASA's National Advisory Council. Dr. Zoloth's publications include *Health Care and the Ethics of Encounter: A Jewish Discussion of Social Justice* (1999) and with Dr. Dena Davis, *Notes from a Narrow Ridge: Religion and Bioethics* (1999).